T0172897

Structural Dynamic Systems Computational Techniques and Optimization

Dynamic Analysis and Control Techniques

*Gordon and Breach International Series
in Engineering, Technology and Applied Science*

Volumes 7–15

Edited by Cornelius T. Leondes

Books on **Structural Dynamic Systems Computational
Techniques and Optimization**

Volume 7 *Computer-Aided Design and Engineering*

Volume 8 *Finite Element Analysis (FEA) Techniques*

Volume 9 *Optimization Techniques*

Volume 10 *Reliability and Damage Tolerance*

Volume 11 *Techniques in Buildings and Bridges*

Volume 12 *Seismic Techniques*

Volume 13 *Computational Techniques*

Volume 14 *Dynamic Analysis and Control Techniques*

Volume 15 *Nonlinear Techniques*

Previously published in this series were volumes 1–6 on
**Medical Imaging Systems Techniques and
Applications**

Forthcoming in the *Gordon and Breach International Series
in Engineering, Technology and Applied Science*

Mechatronics

Biomechanical Systems

Computer Aided Design/Computer Aided Engineering
(CAD/CAE)

Expert Systems

This book is part of a series. The publisher will accept continuation orders which may be
cancelled at any time and which provide for automatic billing and shipping of each title in the
series upon publication. Please write for details.

Structural Dynamic Systems Computational Techniques and Optimization

Dynamic Analysis and Control Techniques

Edited by

Cornelius T. Leondes

Professor Emeritus
University of California
at Los Angeles

Gordon and Breach Science Publishers

Australia • Canada • China • France • Germany • India •
Japan • Luxembourg • Malaysia • The Netherlands •
Russia • Singapore • Switzerland

Amsteldijk 166
1st Floor
1079 LH Amsterdam
The Netherlands

British Library Cataloguing in Publication Data

Structural dynamic systems computational techniques and
    optimization : dynamic analysis and control techniques. –
    (Gordon and Breach international series in engineering,
    technology and applied science ; v. 14 — ISSN 1026-0277)
    1. Structural dynamics 2. Structural optimization
    I. Leondes, Cornelius T.
    624.1'7'0285

ISBN 90-5699-658-4

CONTENTS

Series Description and Motivation   vii

Series Preface   ix

Preface   xi

Techniques in Control-Oriented Reduction of Finite Element
Modeling in Structural Dynamic Systems   1
K. H. Yae and D. J. Inman

Techniques in Active Dynamic Structural Control to Optimize
Structural Controllers and Structural Parameters   55
Mark J. Schulz and Daniel J. Inman

Nonlinear Modal Control Techniques and Applications
in Structural Dynamic Systems   107
J. C. Slater and Greg Agnes

Superelement Modeling of Vehicle Dynamic Structural Systems   169
O. P. Agrawal, K. J. Danhof and R. Kumar

Robust Control Design of a High Performance Flexible
Electro-Mechanical System   211
Y. Chait, M. Steinbuch and M. Soo

A Component Modes Damping Assignment Methodology
for Articulated Multi-Flexible Body Systems   241
Allan Y. Lee

SERIES DESCRIPTION AND MOTIVATION

Many aspects of explosively growing technology are difficult or essentially impossible for one author to treat in an adequately comprehensive manner. Spectacular technological growth is made stunningly manifest by any number of examples, but, just to note one here, the Intel 486 IBM-compatible PC was first introduced in late 1989. At that time the price of this PC was in the $10,000 range and it was thought to be much too powerful for widespread use. By early 1992, a little more than two years later, the price had dropped to $1,000 and it was felt that much more power was needed, leading directly to the Pentium IBM-compatible PC. A similar price reduction pattern has already been projected for the Pentium computer, and, in view of the recent history of the 486, it is difficult to suggest that the same "power hungry" pattern will not occur again in a similar time span. The Pentium is presently planned as a 1,000-MHz processor to be called the Flagstaff in the year 2000. The CD-ROM is presently evolving to the DVD (Digital Versatile Disk) with data storage capability of a greater order of magnitude. A DVD-ROM can hold a database of all the phone numbers and addresses in the United States, which would normally require multiple CD-ROMs. And the DVD format has room to grow. In any event, these examples and their clear implications with respect to the many application-oriented issues in diverse fields of engineering, technology and applied science and their continuing advances make it obvious that this series will fill an essential role in numerous ways for individuals and organizations.

Areas of major significance will be defined and world-class co-authors identified as contributors for essential volumes in respective areas. These areas will be determined by criteria including:

1. Will volumes fill important textbook voids in respective areas?

2. In some cases, a "time void" for an important area will clearly suggest the need for a volume. For example, the important area of Expert Systems might have a textbook void of several years that "requires" an important new volume.

3. Are these technology areas that simply cannot sensibly be treated comprehensively by a single author or even several co-authors?

Examples of areas requiring important volumes will be carefully defined and structured and might include, as the case arises, volumes in:

1. Medical imaging

2. Structural dynamic systems

3. Mechatronics

4. Biomechanical systems

5. Computer aided design/computer aided engineering (CAD/CAE)

6. Expert systems (again, depending on a possible significant time void)

7. Computers in medicine

8. Data base and data communication networks

9. Computer imaging.

One of the most important aspects of this series will be that, despite rapid advances in technology, respective volumes will be defined and structured to constitute works of indefinite or "lasting" reference interest.

SERIES PREFACE

The first industrial revolution, with its roots in James Watt's steam engine and its various applications to modes of transportation, manufacturing and other areas, introduced to mankind novel ways of working and living, thus becoming one of the chief determinants of our present way of life.

The second industrial revolution, with its roots in modern computer technology and integrated electronics technology — particularly VLSI (Very Large Scale Integrated) electronics technology, has also resulted in advances of enormous significance in all areas of modern activity, with great economic impact as well.

Some of the areas of modern activity created by this revolution are: medical imaging, structural dynamic systems, mechatronics, biomechanics, computer-aided and integrated manufacturing systems, applications of expert and knowledge-based systems, and so on. Documentation of these areas well exceeds the capabilities of any one or even several individuals, and it is quite evident that single-volume treatments — whose intent would be to provide practitioners with useful reference sources — while useful, would generally be rather limited.

It is the intent of this series to provide comprehensive multi-volume treatments of areas of significant importance, both the above-mentioned and others. In all cases, contributors to these volumes will be individuals who have made notable contributions in their respective fields. Every attempt will be made to make each book self-contained, thus enhancing its usefulness to practitioners in a specific area or related areas. Each multi-volume treatment will constitute a well-integrated but distinctly titled set of volumes. In summary, it is the goal of the respective sets of volumes in this series to provide an essential service to the many individuals on the international scene who are deeply involved in contributing to significant advances in the second industrial revolution.

PREFACE

Structural Dynamic Systems Computational Techniques and Optimization

Dynamic Analysis and Control Techniques

There are various techniques available to optimize either structural parameters or structural controllers, but there are not many techniques that can simultaneously optimize both. The advantage of integrating structural and controller optimization problems is that their interaction is taken into account in the design process and a more efficient overall design (lower control force/lighter weight) can be achieved while multi-disciplinary design optimization also can be performed. The down side is that the combined optimization problem is more difficult to formulate and solve, and computations are increased. This volume is a comprehensive treatment of dynamic analysis and control techniques in structural dynamic systems and the wide variety of issues and techniques that fall within this broad area, including the interactions between structural control systems and structural system parameters.

This is the eighth volume in the set of 9 volumes on structural dynamic systems. Subjects treated are:

1. Computer-Aided Design and Engineering

2. Finite Element Analysis (FEA) Techniques

3. Optimization Techniques

4. Reliability and Damage Tolerance

5. Techniques in Buildings and Bridges

6. Seismic Techniques

7. Computational Techniques

8. Dynamic Analysis and Control Techniques

9. Nonlinear Techniques.

The first contribution to this volume tells us the finite element method has been applied extensively in analyzing dynamic structures. Although versatile, it generates a large number of degrees of freedom for desired accuracy. It has therefore employed condensation methods, such as Guyan reduction, to reduce the size of mass and stiffness matrices by eliminating insignificant displacements at the node points. The resulting model is, however, still large enough to warrant further reduction from the input/output point of view. In recent years experimental modal analysis techniques have been improved considerably, thanks to advances of computer technology in measurement and estimation. Currently, the modal information from experiments is used not only to verify the analytical model but also to compare and modify it on the basis of orthogonality or weighted orthogonality relationships. Such model modification is necessary because the experimental modal model is of lower order than the finite element analysis model. This chapter is a comprehensive treatment of effective methods in reduction of finite element models of structural dynamic systems as they relate to control of such systems. This is an area of broad and significant applications and numerous illustrative examples add greatly to the discussion.

In chapter 2, Schulz and Inman relate that although there are many accepted techniques to design controllers and optimize structures separately, there is no universal technique to simultaneously design structures and structural controllers. This is because optimal control algorithms are difficult to couple to a structural optimization, and adjusting the controller and structural variables simultaneously using multi-variable optimization is computationally very intensive. Although the global optimum solution is difficult to find when adjusting a large number of design variables, it is becoming necessary to use multi-disciplinary design optimization to meet increasingly stringent design requirements. This chapter treats the issues involved in this broadly complex and substantively significant area.

According to Slater and Agnes (chapter 3), the control of multiple-degree-of-freedom nonlinear systems is perhaps the most challenging structural control problem engineers face today. Nonlinear oscillatory systems are quite common (i.e., robotics, slewing, large strain vibrations, vibration of bimodulus composite materials, the motion of a swinging spring, vibrations of shells and composite plates) and the nonlinear control of these systems is becoming more possible with the advent of ever faster microprocessors. Popular methods used today include Lyapunov design, adaptive control, linearizing control, sliding mode control, fuzzy logic and neural networks. Lyapunov-based control design can be unwieldy for nonlinear systems with many degrees of freedom. Linearizing control is likely to be wasteful of energy, since the linearizing part of the control design does nothing to improve actual performance. Often it is not feasible, since the actuators and sensors must be placed in locations where they are able to counteract the nonlinear effects, which is usually possible only for nonlinear effects in actuators. Fuzzy logic and neural network controls assume an ambiguity that does not necessarily exist, and thus can add unwarranted complexity to the control problem. One concept that has not been fully exploited with respect to nonlinear systems is that of nonlinear modal control. This contribution is a comprehensive discussion of methods for treating nonlinear oscillatory structural dynamic systems that are characteristic of many significant applications. Because of their significant effectiveness in treating this problem, nonlinear modal control techniques are examined in depth.

In chapter 4, by Agrawal et al., we learn that dynamic simulation and control of ground vehicles require access to high-speed computational engines and efficient analytical formulations and numerical algorithms, especially when real-time simulation, modeling of flexibility and crash-worthiness, man-in-the-loop, and instant graphic visualizations are considered. Significant progress has been made in the areas of formulations and algorithms, and computational hardware tools. Current formulations are several times faster than those developed 20 years ago due to extensive progress made in the area of multi-body dynamics. Recent emergence of high-speed computational tools and environments such as vector, multi- and massively parallel processors, multi-threading operating systems, computer networking, and high-speed graphics engines have considerably changed the way vehicle systems are analyzed. Vector, multi- and massively parallel processor machines allow execution of several instructions in parallel. Networking allows computation of different modules associated with a simulation in parallel on different remote machines. Usage of concurrent programming paradigms based on user level threads, for example, as implemented in the Sun OS/Solaris operating system, do help exploit the newer and faster multi-processor machines to various degrees depending on how much the uni-processor software base has been modified. High-speed graphics engines allow an instant display of simulation results; all this has increased the computational speed.

These advances in computer software and hardware technologies have motivated investigators to develop new vector-parallel algorithms for vehicle dynamics; and the real-time simulation and control including real-time graphical visualization and man-in-the-loop are now a reality. Although progress has been made in the area of parallel algorithms for real-time simulation of vehicle systems, much remains to be done. Development of efficient parallel algorithms for the above tasks relies on two major efforts: the efficient utilization of current software and hardware technologies, and the efficient analytical formulations for the systems. The distribution of computational load among all processors should be such that processor-idle-time is minimal, and analytical formulation should be such that it leads to computation of the minimum number of terms.

A common problem encountered in control of large flexible space structures is complexity of dynamics, namely, a large number of lightly damped
flexible modes in the systems. To meet stringent performance specifications,
such structures must be actively controlled (chapter 5) so as to minimize
contribution of flexible modes in transient response. Control design requires
knowledge of plant dynamics; and the more accurate the model, the better
the achievable controlled performance — with lower costs. Analytic modeling
of such structures is an inexact science, at the least, while obtaining models
using extensive experimentation is often not feasible. Therefore, achieving
robustness for parasitic dynamics becomes critical for successful design of
high-performance control systems. In contrast to such systems, small actuator drives such as compact-disc and hard-disc drive mechanisms are easy
to access with experiments. In these high-volume electronics applications,
the problem is to allow for tolerances during production, so as to decrease
production costs. Hence, again we see the motivation to design robust
performance controllers. As do large flexible structures, the dynamics includes flexible modes. Although model identification is more feasible,
uncertainty modeling is still necessary to account for changes from product
to product. Such compact-disc and hard-disc drives are an essential element
of modern computer technology, and thus the issues and techniques presented in this contribution are of substantial significance as a major class
of structural dynamic systems. Illustrative examples are included.

The final chapter to this volume discusses multi-flexible body systems,
which are assemblages of rigid and flexible bodies where each body can
rotate and/or translate relative to the others. Examples are spacecraft, robotic
manipulators, helicopters and other systems. The equations of motion of
these systems are typically very complex. As a result, a multi-body simulation package must be used to determine time responses of these systems,
when they are subjected to external dynamic stimuli. To this end, one must
supply appropriate models for both the rigid-body and flexible components
involved. Chapter 6 develops methodology for this complex problem and
illustrates the effectiveness of techniques using a high-order model of the
Galileo spacecraft.

This volume on dynamic analysis and control techniques in structural
dynamic systems reveals the effectiveness and essential significance of
techniques available — and with further development — the essential role
they will play in the future. The authors are all to be highly commended for
their splendid contributions; these papers will provide a significant and
unique reference source for students, research workers, practitioners, computer scientists and others on the international scene for years to come.

1 TECHNIQUES IN CONTROL-ORIENTED REDUCTION OF FINITE ELEMENT MODELING IN STRUCTURAL DYNAMICS SYSTEMS

K.H. YAE[1] and D.J. INMAN[2]

[1] *Samsung Advanced Institute of Technology, P.O. Box 111, Suwon, 440–600, S. Korea*
[2] *Virginia Polytechnic Institute and State University, USA*

1.1. INTRODUCTION

The finite element method has been applied extensively in analyzing dynamic structures. Although versatile, it generates a large number of degrees of freedom for desired accuracy. It has therefore employed condensation methods, such as Guyan reduction, to reduce the size of the mass and stiffness matrices by eliminating insignificant displacements at the node points. The resulting model is, however, still large enough to warrant further reduction from the input/output point of view.

In the recent years experimental modal analysis techniques have been improved considerably, thanks to advances of computer technology in measurement and estimation. Currently, the modal information from experiments is used not only to verify the analytical model but also to compare and modify the analytical model on the basis of orthogonality or weighted orthogonality relationship. Such model modification is necessary because the experimental modal model is of order lower than the finite element analysis model. In this chapter a model modification technique is proposed, which can be applied to a finite element model. This technique is based upon the experimental modal model and the eigenvalue problem.

Many model reduction methods for linear dynamical systems in the state space form have been proposed in the control literature. Lately, a method based on the measures of controllability and observability has been suggested by Moore (1981). In this approach a system in the state space form is transformed into an equivalent, "internally balanced" system. In the internally–balanced coordinates all the states are rearranged according to the measures of their controllability and observability. Thus a reduced–order model can be obtained after the least controllable and the least observable states have been removed. The resulting, reduced model is expressed in a coordinate system different from the original; that is, it is in the *balanced* coordinate system. A number of states that are removed in the balanced coordinate system can be expressed as linear combinations of the original states, which forces the same number of the original states linearly dependent. Hence, these states can be removed, with model reduction error, from the original list of state variables and the remaining states represent the physical meaning (more or less) same as before.

One of the important and basic conditions often used in dynamic modeling is that experimental models satisfy the theoretical requirement of weighted orthogonality with respect to the mass and stiffness matrices. Such a requirement can only be satisfied by assuming no, or proportional damping. For structures with non–proportional damping the orthogonality condition is applied in various ways, because of the non–unique conversion from a physical model into a state–space form and because of infinite number of complex modal vectors related through linear transformations. The eigenvalue problem is, however, uniquely defined in the state–space form as well as in the physical model, which makes it possible to modify the reduced model with minimum change in the eigenvalues (natural frequencies) and eigenvectors (mode shape vectors) identified by modal testing and estimation.

The final reduced model reflects not only the original finite element model but also the experimental modal data. Its order is smaller than the original finite element model, making the reduced model more agreeable to work with from the control theorists' point of view.

The rest of the sections are organized as follows.

In Section 2 model reduction methods are briefly surveyed in the two different areas of finite element analysis and control theory.

In Section 3 the main results are presented. The model reduction by the balancing method and the model modification are discussed. The estimation technique for a non–proportional damping matrix is presented. The mathematical details of the results in this section are given in Appendices A, B, and C.

In Section 4 the numerical algorithm is summarized for programming.

In Section 5 the main results are applied to a simple structure and the procedure is illustrated numerically in detail.

Section 6 provides a summary of the results and remarks on continuing research.

1.2. PREVIOUS WORK IN MODEL REDUCTION

Several model reduction methods are briefly surveyed in the areas of finite element analysis and control theory. The reduced–order modeling in finite element analysis, in the case of a condensation process, actually removes some of the physical coordinates of the model. On the other hand, the model reduction in control theory results in a smaller model in terms of input/output characteristics, in which the reduced model is expressed in a coordinate system different from the original.

1.2.1. Coordinate Reduction in FEM

Rayleigh–Ritz principle with high–speed digital computers has generated a powerful method of analysis, that is, the finite element method. Application of the finite element method generally results in stiffness and mass matrices of high order, which in turn can place severe strains on various computational algorithms. Even ordinary operations such as matrix inversion can become a problem for high–order matrices. One approach to the problems of high dimensionality is to reduce the order of the system by eliminating unnecessary or unwanted variables, a process known as *condensation*.

1.2.1.1. Static Condensation

It is common practice in finite element analysis to use an *inconsistent* mass matrix in the form of a diagonalized, lumped mass matrix. When the nodal displacements include rotation, as in bending vibration, the associated entries in the mass matrix would have to be moments of inertia. By lumped mass we generally imply point masses and seldom include moments of inertia, unless there is a good reason to believe that rotary inertia effects are important. Hence, the diagonal entries associated with rotational displacements are generally zero. The fact that certain diagonal entries in the lumped mass matrix are zero is an indication that the corresponding displacements are not vital to the solution and can be eliminated from the problem formulation. The elimination process is known as *static condensation* and its net result is to reduce the order of the eigenvalue problem. It is worth pointing out that static condensation involves no approximation.

Let us consider the eigenvalue problem

$$K\mathbf{u} = \lambda M\mathbf{u} \tag{1}$$

where K is the symmetric stiffness matrix, M the diagonal lumped mass matrix and $\mathbf{u}$ the nodal displacement vector. Then, assuming that there are certain zero diagonal entries, we can arrange the mass matrix M in a way that all the zero diagonal entries form a null submatrix. This enables us to partition the eigenvalue problem as follows:

$$\begin{bmatrix} K_{11} & K_{12} \\ K_{21} & K_{22} \end{bmatrix} \begin{bmatrix} \mathbf{u}_1 \\ \mathbf{u}_2 \end{bmatrix} = \lambda \begin{bmatrix} M_{11} & 0 \\ 0 & 0 \end{bmatrix} \begin{bmatrix} \mathbf{u}_1 \\ \mathbf{u}_2 \end{bmatrix} \tag{2}$$

The above equation can be separated into the two equations

$$K_{11}\mathbf{u}_1 + K_{12}\mathbf{u}_2 = \lambda M_{11}\mathbf{u}_1 \tag{3}$$

$$K_{21}\mathbf{u}_1 + K_{22}\mathbf{u}_2 = 0 \tag{4}$$

Solving Equation (4) for $\mathbf{u}_2$ yields,

$$\mathbf{u}_2 = -K_{22}^{-1}K_{21}\mathbf{u}_1 \tag{5}$$

and substituting Equation (5) into Equation (3) yields the *condensed eigenvalue problem*

$$\left(K_{11} - K_{12}K_{22}^{-1}K_{21} \right) \mathbf{u}_1 = \lambda M_{11}\mathbf{u}_1 \tag{6}$$

EXAMPLE: Consider a cantilever beam with two node points: one in the middle and the other at the tip, denoted by node 1 and node 2, respectively. Each node point has two degrees of freedom: translational displacement (x_i) and rotational displacement (θ_i). Here rotational displacements are considered as much less important than the translational. Physical dimensions and parameters are given as follows:

Length $= 2L$
Cross–section area $= A$
Moment of inertia $= I$
Density of the material $= \rho$
Young's modulus $= E$

The stiffness matrix is, then, given by

$$K = \frac{EI}{L^3} \begin{bmatrix} 24 & -12 & 0 & 6L \\ -12 & 12 & -6L & -6L \\ 0 & -6L & 8L^2 & 2L^2 \\ 6L & -6L & 2L^2 & 4L^2 \end{bmatrix} \tag{7}$$

with the corresponding nodal displacement vector

$$\begin{bmatrix} x_1 \\ x_2 \\ \theta_1 \\ \theta_2 \end{bmatrix} \tag{8}$$

The inconsistent mass matrix M_{11} is given by

$$M_{11} = \frac{\rho A L}{420} \begin{bmatrix} 420 & 0 \\ 0 & 210 \end{bmatrix} \tag{9}$$

Thus, the condensed stiffness matrix is obtained as

$$(K_{11} - K_{12} K_{22}^{-1} K_{21}) = \frac{EI}{L^3} \begin{bmatrix} 13.71 & -4.29 \\ -4.29 & 1.71 \end{bmatrix} \tag{10}$$

1.2.1.2. Mass Condensation

The inertia force components associated with some of the displacements are sometimes known to be smaller than those associated with others, so that these displacements can be regarded less important in an overall solution, even though all displacements play an equal part in computation of the eigenvalue problem. Hence, the question arises as to whether it is possible to eliminate displacements of lesser importance from the problem formulation, without affecting the results too much. The *mass condensation* is, in essence, the sacrifice of accuracy for a reduction in computational effort. The references for this Section are Irons (1965), Guyan (1965), Kidder (1973), and Meirovitch (1980).

Let us write the potential and the kinetic energy of a system in matrix form

$$V = \frac{1}{2} \mathbf{q}^T K \mathbf{q}$$

$$T = \frac{1}{2} \dot{\mathbf{q}}^T M \dot{\mathbf{q}} \tag{11}$$

and divide the displacement vector $\mathbf{q}$ into the *significant* displacement vector $\mathbf{q}_1$ and *insignificant* displacement vector $\mathbf{q}_2$, or

$$\mathbf{q} = \begin{bmatrix} \mathbf{q}_1 \\ \mathbf{q}_2 \end{bmatrix} \tag{12}$$

Then, the stiffness matrix K and the mass matrix M can be partitioned accordingly. The potential and the kinetic energy can be written in terms of the partitioned matrices and vectors,

$$
\begin{aligned}
V &= \frac{1}{2} \begin{bmatrix} \mathbf{q}_1 \\ \mathbf{q}_2 \end{bmatrix}^T \begin{bmatrix} K_{11} & K_{12} \\ K_{21} & K_{22} \end{bmatrix} \begin{bmatrix} \mathbf{q}_1 \\ \mathbf{q}_2 \end{bmatrix} \\
T &= \frac{1}{2} \begin{bmatrix} \dot{\mathbf{q}}_1 \\ \dot{\mathbf{q}}_2 \end{bmatrix}^T \begin{bmatrix} M_{11} & M_{12} \\ M_{21} & M_{22} \end{bmatrix} \begin{bmatrix} \dot{\mathbf{q}}_1 \\ \dot{\mathbf{q}}_2 \end{bmatrix}
\end{aligned}
\tag{13}
$$

where $K_{21} = K_{12}^T$ and $M_{21} = M_{12}^T$. The condition that there be no applied force in the direction of insignificant displacements can be written symbolically in form

$$\frac{\partial V}{\partial \mathbf{q}_2} = K_{21}\mathbf{q}_1 + K_{22}\mathbf{q}_2 = 0 \tag{14}$$

where the equation implies static equilibrium in the direction of insignificant displacements. Solving it for $\mathbf{q}_2$, we obtain

$$\mathbf{q}_2 = -K_{22}^{-1} K_{21}\mathbf{q}_1 \tag{15}$$

which can be used to eliminate $\mathbf{q}_2$ from the problem formulation. The original displacement vector $\mathbf{q}$ can be expressed in terms of the significant displacement vector $\mathbf{q}_1$ in form

$$\mathbf{q} = C\mathbf{q}_1 \tag{16}$$

where C is a rectangular constraint matrix

$$C = \begin{bmatrix} I \\ -K_{22}^{-1} K_{21} \end{bmatrix} \tag{17}$$

in which I is a unit matrix of the same order as the dimension of $\mathbf{q}_1$. So the potential and kinetic energy can be written as

$$
\begin{aligned}
V &= \frac{1}{2} \mathbf{q}_1^T K_1 \mathbf{q}_1 \\
T &= \frac{1}{2} \dot{\mathbf{q}}_1^T M_1 \dot{\mathbf{q}}_1
\end{aligned}
\tag{18}
$$

where the reduced stiffness and mass matrices are simply

$$K_1 = C^T K C = K_{11} - K_{12} K_{22}^{-1} K_{21}$$
$$M_1 = C^T M C = M_{11} - K_{21}^T K_{22}^{-1} M_{21} - M_{12} K_{22}^{-1} K_{21} \qquad (19)$$
$$+ K_{21}^T K_{22}^{-1} M_{22} K_{22}^{-1} K_{21}$$

The matrix M_1 is generally known as the *condensed mass matrix*.

Mass condensation is similar to static condensation. The difference is that whereas static condensation involves no approximation, mass condensation does involve some approximation. In order to see what is being sacrificed as a result of the condensation process, let us consider the complete eigenvalue problem, which can be separated into

$$K_{11}\mathbf{q}_1 + K_{12}\mathbf{q}_2 = \lambda(M_{11}\mathbf{q}_1 + M_{12}\mathbf{q}_2) \qquad (20)$$

$$K_{21}\mathbf{q}_1 + K_{22}\mathbf{q}_2 = \lambda(M_{21}\mathbf{q}_1 + M_{22}\mathbf{q}_2) \qquad (21)$$

By substituting Equation (21) for $\mathbf{q}_1$, we have

$$\mathbf{q}_2 = (K_{22} - \lambda M_{22})^{-1}(\lambda M_{21} - K_{21}) \qquad (22)$$

so that, introducing Equation (22) into Equations (20) and (21), we obtain

$$(K_{11} - K_{12} K_{22}^{-1} K_{21})\mathbf{q}_1$$
$$= \lambda(M_{11} - K_{12} K_{22}^{-1} M_{21} - M_{12} K_{22}^{-1} K_{21} + K_{12} K_{22}^{-1} M_{22} K_{22}^{-1} K_{21})\mathbf{q}_1$$
$$+ \lambda^2(M_{12} K_{22}^{-1} M_{21} - M_{12} K_{22}^{-1} M_{22} K_{22}^{-1} K_{21} - K_{12} K_{22}^{-1} M_{22} K_{22}^{-1} M_{21}$$
$$+ K_{12} K_{22}^{-1} M_{22} K_{22}^{-1} M_{22} K_{22}^{-1} K_{21})\mathbf{q}_1$$
$$+ \; higher \; order \; terms$$

$$(23)$$

Examining Equations (19) and (23), we conclude that the mass condensation process is equivalent to ignoring second– and higher–order terms in λ, which can be justified if the elements of the coefficients of λ^2, λ^3, ... are significantly smaller than those of the coefficient of λ. In order for this to be true, the entries of M_{12} and M_{22} should be much smaller than the entries of K_{12} and K_{22}. Physically, this implies that insignificant displacements should be chosen from areas of high stiffness and low mass.

EXAMPLE: The model in the previous example is used here to demonstrate the mass condensation. Translational displacements are considered as *significant* and rotational displacements as *insignificant*. The *consistent* mass matrix is found as

$$M = \frac{\rho A L}{420} \begin{bmatrix} 312 & 54 & 0 & -13L \\ 54 & 156 & 13L & -22L \\ 0 & 13L & 8L^2 & -3L^2 \\ -13L & -22L & -3L^2 & 4L^2 \end{bmatrix} \qquad (24)$$

with the corresponding nodal displacement vector

$$\begin{bmatrix} x_1 \\ x_2 \\ \theta_1 \\ \theta_2 \end{bmatrix} \tag{25}$$

Then the condensed mass matrix is obtained as

$$M_1 = \frac{\rho AL}{420} \begin{bmatrix} 374.2 & -1.65 \\ -1.65 & 228.5 \end{bmatrix} \tag{26}$$

1.2.2. Model Reduction in the Control Literature

There has been much research in the approximation of complex physical models. One only has to examine the comprehensive list of references compiled by Genesio and Milanese (1976) to appreciate this fact. One problem that has received a particular attention is the approximation of a large–order linear dynamical systems by dynamical systems having fewer state variables. The model reduction should aim not only at obtaining a good approximation of response to particular inputs but at using a reduced–order model in feedback control, i. e., the application of model reduction to control design.

According to Hickin and Sinha (1980) the techniques of approximating large–scale linear time–invariant dynamical systems by low–order models may be classified in two main groups, those which preserve eigenvalues (aggregated models) and those which preserve part of Taylor series of the system transfer function (partial realization). Furthermore, these two methods of reduction may be simultaneously applied. An important question that still remains unanswered is which of the eigenvalues of the original system should be retained in the reduced–order model.

The following four methods are outlined here for their importance as recent developments in the area of reduced–order modeling.

1.2.2.1. Aggregation Method

The model is derived by "aggregating" the original system state vector into a lower–dimensional vector. The concept of aggregation, a generalization of projection, is related to state vector partitioning and is useful not only in building a model of reduced dimension, but also in unifying several topics in control theory such as the regulator with incomplete state feedback, characteristic value computations, model controls, and bounds on the solution of the matrix Riccati equations.

Consider two linear dynamic systems S_1 and S_2, where $\dim(S_1) = n$, $\dim(S_2) = l$, and $n \gg l$. The state vectors of these two systems, denoted by $\mathbf{x}$ and $\mathbf{z}$, respectively, are to satisfy the relationship under certain conditions $\mathbf{z} = C\mathbf{x}$ where C is $l \times n$ constant matrix.

There are at least two, not necessarily mutually exclusive, viewpoints on the way such a relationship arises.

(1) S_1 is regarded as a given system and S_2 as an observer of S_1 (Luenberger, 1963). The linear transformation C is to be chosen, subject to certain constraints, such that the state vector $\mathbf{z}$, together with originally available measurement $\mathbf{y}$ on S_1, is used to reconstruct $\mathbf{x}$ exactly or approximately. Loosely speaking, the dynamic structure of S_2 is chosen to reflect only that part of the dynamics of S_1, which is not carried by the independent information contained in $\mathbf{y}$.

(2) S_2 is regarded as a model of S_1. The dynamic structure of S_2 is to reflect the significant portion of the dynamics of S_1, not to complement the information carried by $\mathbf{y}$. S_1 is the description of a physical or nonphysical object according to some classification of variables, and S_2 is the description of the same object using a coarser grid of classification, hence, of lesser dimension. S_2 is called *an aggregated model* of S_1.

One of the important questions to be examined in controlling S_1 by control law derived for S_2 is the performance degradation suffered by S_1 and the stability of S_1.

Designers have some freedom in choosing the aggregation matrix C subject to constraints imposed by specific problems. For example, for certain C, S_2 becomes a projection of S_1 into the l–dimensional state space, retaining the l most significant characteristic values and characteristic vectors of S_1. The model obtained by such a projection method is a special case of models constructed by aggregation.

Consider a continuous–time dynamic system described by the equation

$$S_1 : \quad \dot{\mathbf{x}} = A\mathbf{x} + B\mathbf{u} \qquad (27)$$

where $B \in \mathbf{R}^{n \times p}$ and where $\mathbf{x}$ and $\mathbf{u}$ are the state and control vectors of appropriate dimensions, respectively. Let an l–dimensional vector $\mathbf{z}$ be the *aggregated state vector*,

$$\mathbf{z}(t) = C\mathbf{x}(t) \qquad (28)$$

where $C \in \mathbf{R}^{l \times n}$, $\mathrm{rank}[C] = l$, and $l < n$. The aggregated model S_2 is expressed as

$$S_2 : \quad \dot{\mathbf{z}} = F\mathbf{z} + G\mathbf{u}, \quad \mathbf{z}(0) = C\mathbf{x}(0) \qquad (29)$$

F and G are related to A and B by

$$FC = CA$$
$$G = CB \tag{30}$$

If A and C satisfy the matrix equation

$$CA = CAC'(CC')^{-1}C \tag{31}$$

then F is given by

$$F = CAC'(CC')^{-1}. \tag{32}$$

The matrix C is referred to as *the aggregation matrix* and F as *the aggregated matrix* or *the aggregation of* A.

C is the primary design parameter in constructing S_2. The choice of C is to be made in such a way that error in modeling the original system S_1 by S_2 is minimized, in some sense taking into account the performance index for S_1 (Aoki, 1968).

1.2.2.2. Modal Cost Analysis

Skelton and Hughes (1980) derived modal cost analysis for linear matrix–second–order systems. The system given by the matrix–second–order equation is converted to first–order (state space) form. Through a linear transformation the first–order equations are uncoupled into Jordan modal coordinates. The decomposition of a quadratic cost index into the sum of contributions from each modal coordinate can be used to rank the importance of modes in the model and control problem. The contribution of each mode to the cost is called the "modal cost" and may be used as a modal truncation criterion. Modal costs are given in detail in Appendix B of Skelton and Hughes (1980).

The system of interest is given by

$$\dot{\mathbf{x}} = A\mathbf{x} + B\mathbf{u} \tag{33}$$

$$\mathcal{E}\{\mathbf{x}(0)\} = 0; \quad \mathcal{E}\{\mathbf{x}(0)\mathbf{x}(0)^T\} = \Sigma \tag{34}$$

$$\mathbf{u} = G\mathbf{x}, \quad \mathbf{y} = P\mathbf{x} \tag{35}$$

$$V = \mathcal{E} \int_0^\infty (\mathbf{y}^T Q\mathbf{y} + \mathbf{u}^T R\mathbf{u})dt \tag{36}$$

where $\mathbf{x} \in \mathbf{R}^n$, $\mathbf{u} \in \mathbf{R}^{n_c}$, $\mathbf{y} \in \mathbf{R}^{n_o}$, and $\mathcal{E}\{\cdot\}$ is the expectation operator. Further let $\mathbf{e}_i$ and $\mathbf{f}_i$ be the the right and left eigenvectors of $A + BG$ in the sense that, if

$$E = [\mathbf{e}_1 \quad \cdots \quad \mathbf{e}_n] \tag{37}$$

and

$$F = \begin{bmatrix} \mathbf{f}_1^* \\ \vdots \\ \mathbf{f}_n^* \end{bmatrix} \tag{38}$$

then

$$EF = I \tag{39}$$

$$(A + BG)E = E\Lambda; \quad F(A + BG) = \Lambda F \tag{40}$$

where $\Lambda = \mathrm{diag}\{\lambda_1, \dots, \lambda_n\}$ and λ_i's are the eigenvalues of $A + BG$ (closed–loop poles) and where the superscript $*$ means complex conjugate transpose. E is *the modal matrix* for $A + BG$.

The following definitions are helpful

$$\begin{aligned} \mathbf{o}_i &= P\mathbf{e}_i \\ \mathbf{g}_i &= G\mathbf{e}_i \end{aligned} \tag{41}$$

where $\mathbf{o}_i$ and $\mathbf{g}_i$ are called the (closed–loop) *modal observability vector* and the *modal gain* for mode i, respectively. The cost function V can be written

$$\begin{aligned} V &= \mathcal{E} \int_0^\infty (\mathbf{y}^T Q\mathbf{y} + \mathbf{u}^T R\mathbf{u})dt \\ &= \mathrm{tr} \int_0^\infty \left[Q \cdot \mathcal{E}\{\mathbf{y}\mathbf{y}^T\} + R \cdot \mathcal{E}\{\mathbf{u}\mathbf{u}^T\} \right] dt \\ &= \sum_{i=1}^n V_i \end{aligned} \tag{42}$$

$$V_i = -\sum_{i=1}^n \frac{(\mathbf{o}_i^* Q\mathbf{o}_j + \mathbf{g}_i^* R\mathbf{g}_j)\mathbf{f}_j^* \Sigma \mathbf{f}_i}{\bar{\lambda}_i + \lambda_j} \tag{43}$$

with the assumption $\mathrm{Re}\{\bar{\lambda}_i + \lambda_j\} < 0$. V_i is called *the modal cost* for mode i.

The primary obstacle in applying closed–loop modal cost analysis is the determination of the initial control gains before model reduction begins. A two–phase procedure is suggested to circumvent this difficulty. The open–loop, $G = 0$, modal cost analysis may first be used to reduce the model to an order that is tractable for computation of control gains. In the second phase the closed–loop modal cost analysis can then be applied to further reduce the controller equations. This procudure identifies and deletes from control policy those modes of the smallest contribution to the control problem.

For the synthesis of the reduced–order control law, the use of state estimation is suggested by Skelton and Hughes (1980).

The references for this section are Skelton and Yousuff (1983); Skelton, Hughes, and Hablani (1982); Skelton (1980); Skelton and Hughes (1980).

1.2.2.3. *Internal Balancing*

This method is based on measures of controllability and observability (Moore, 1981). A natural way to decrease the order of a model is to delete everything but the controllable and observable part. The problem is that any system, in some sense, is generically controllable and observable. Therefore, one would like to be able to measure the "degree of controllability and observability" of different subspaces of the state space. The most controllable and observable part could then be used as a low–order approximation for the model.

The measures of controllability and observability is quantified by the controllability and observability grammians. Since the grammians are not invariant under coordinate transformations, it is shown that there exists a coordinate system in which the two grammians are equal and diagonal. The corresponding system representation is called *balanced*. A reduced–order model can be obtained from the balanced representation by deleting the least controllable and, therefore, least observable part.

The references for this section are Gawronski and Natke (1986, 1987(b)); Moore (1981); Pernebo and Silverman (1982); Shokoohi, Silverman, and Van Dooren (1983); Viswanathan, Longman, and Likins (1984).

1.2.2.4. *Optimal Projection*

Hyland and Bernstein (1985) have derived first–order necessary conditions for a quadratically–optimal reduced–order model of a linear time–invariant system. It is shown how the complex optimality conditions of Wilson (1970) can be transformed without loss of generality into much simpler and more tractable forms. The transformation is facilitated by exploiting the presence of an oblique (i.e. nonorthogonal) projection, which was not recognized in Wilson (1970) and which arises as a direct consequence of optimality. The resulting "optimal projection equations" constitute a coupled system of two $n \times n$ modified Lyapunov equations whose solutions are given by a pair of rank–n_m controllability and observability *pseudogrammians*. The highly–structured form of these equations gives crucial insight into the set of local extrema satisfying the first–order necessary conditions.

The optimal projection method also demonstrates the quadratic extremality and nonoptimality of the balancing method of Moore (1981). A component–

cost analysis of the model–error criterion, similar to the approach of Skelton (1980), is utilized at each iteration to direct the algorithm to the global minimum.

Optimal Model–Reduction Problem: Given the controllable and observable system

$$\dot{\mathbf{x}} = A\mathbf{x} + B\mathbf{u}$$
$$\mathbf{y} = C\mathbf{x} \tag{44}$$

find a reduced–order model

$$\dot{\mathbf{x}}_m = A_m\mathbf{x}_m + B_m\mathbf{u}$$
$$\mathbf{y}_m = C_m\mathbf{x}_m \tag{45}$$

which minimizes the quadratic model–reduction criterion

$$J(A_m, B_m, C_m) = \lim_{t \to \infty} \mathcal{E}\left[(\mathbf{y} - \mathbf{y}_m)^T R(\mathbf{y} - \mathbf{y}_m)\right] \tag{46}$$

where the input $\mathbf{u}(t)$ is white noise with positive–definite intensity V. In order to guarantee that J is finite, it is assumed that A is stable, and the set of admissible reduced–order models is

$$\mathcal{A}_+ = \Big\{(A_m, B_m, C_m) : \quad A_m \text{ is stable}, (A_m, B_m) \text{ is controllable and}$$
$$(A_m, C_m) \text{ is observable}\Big\} \tag{47}$$

Main Theorem: Suppose $(A_m, B_m, C_m) \in \mathcal{A}_+$ solves the optimal model–reduction problem. Then there exist nonnegative definite matrices $\hat{Q}, \hat{P} \in \mathbf{R}^{n \times n}$ such that $A_m, B_m,$ and C_m are given by

$$A_m = \Gamma A G^T$$
$$B_m = \Gamma B \tag{48}$$
$$C_m = C G^T$$

where (G, M, Γ) is a factorization of $\hat{Q}\hat{P}$ in the following way

$$\hat{Q}\hat{P} = \Phi \begin{bmatrix} \Lambda & 0 \\ 0 & 0 \end{bmatrix} \Phi^{-1}$$
$$= \Phi \begin{bmatrix} S \\ 0 \end{bmatrix} [S^{-1}\Lambda S][S^{-1} \quad 0]\Phi^{-1} \tag{49}$$
$$= G^T M \Gamma$$

for all $n_m \times n_m$ invertible S,

$$G^T = \Phi \begin{bmatrix} S \\ 0 \end{bmatrix} \quad M = [S^{-1}\Lambda S] \quad \Gamma = [S^{-1} \ \ 0]\Phi^{-1}$$
$$\Gamma G^T = I_{n_m} \tag{50}$$

The numerical algorithm that solves for $\hat{Q}$ and $\hat{P}$ is given in Hyland and Bernstein (1985):

(1) Initialize $\tau^{(0)} = I_n$. Solve for W_c

$$0 = AW_c + W_c A^T + BVB^T \tag{51}$$

(2) Solve for $\hat{Q}^{(k)}$, $\hat{P}^{(k)}$

$$0 = (A - A_1^{(k)})\hat{Q}^{(k)} + \hat{Q}^{(k)}(A - A_1^{(k)})^T + BVB^T$$
$$0 = (A - A_2^{(k)})^T \hat{P}^{(k)} + \hat{P}^{(k)}(A - A_2^{(k)}) + C^T RC \tag{52}$$

where

$$A_1^{(k)} = \tau^{(k)} A \tau_\perp^{(k)}$$
$$A_2^{(k)} = \tau_\perp^{(k)} A \tau^{(k)} \tag{53}$$

(3) Perform the internal balancing on $\hat{Q}^{(k)}$ and $\hat{P}^{(k)}$

$$\Phi^{(k)} \hat{Q}^{(k)} (\Phi^{(k)})^T = (\Phi^{(k)})^{-T} \hat{P}^{(k)} (\Phi^{(k)})^{-1} = \Sigma^{(k)} \tag{54}$$

$$\Sigma^{(k)} = \mathrm{diag}\{\sigma_1^{(k)}, \dots, \sigma_n^{(k)}\}, \quad \sigma_1^{(k)} \geq \sigma_1^{(k)} \geq \cdots \geq \sigma_n^{(k)} \geq 0. \tag{55}$$

(4) If $k > 1$, check for convergence

$$e_k = \left[\frac{\mathrm{tr}(C^T RCW_c) - \mathrm{tr}(C^T RC\tau^{(k)} \hat{Q}^{(k)} \tau^{(k)T})}{\mathrm{tr}(C^T RCW_c)} \right]^{1/2} \tag{56}$$

If $|e_k - e_{k-1}| <$ tolerance, then go to (8); else continue;

(5) Select n_m eigenprojections

$$\prod_{i_r} \left[\hat{Q}^{(k)} \hat{P}^{(k)} \right] = (\Phi^{(k)})^{-1} E_{i_r} \Phi^{(k)} \quad for \ r = 1, \dots, n_m \tag{57}$$

where E_{i_r} = matrix with unity at the (i_r, i_r) position and zeros elsewhere.

(6) Update

$$\tau^{(k+1)} = \sum_{r=1}^{n_m} \prod_{i_r} \left[\hat{\hat{Q}}^{(k)} \hat{\hat{P}}^{(k)} \right] \tag{58}$$

(7) Check for convergence; if not, increment k and return to (2).
(8) Set

$$\hat{Q} = \tau^{(\infty)} \hat{\hat{Q}} (\tau^{(\infty)})^T, \ \hat{P} = (\tau^{(\infty)})^T \hat{\hat{P}} \tau^{(\infty)} \tag{59}$$

It should be noted that

(i) If a solution exists, $\tau^{(k)} \to \tau$, $\hat{\hat{Q}}^{(k)} \to \hat{Q}$ and $\hat{\hat{P}}^{(k)} \to \hat{P}$ as $k \to \infty$.
(ii) $\tau A \tau_\perp \hat{Q} = 0$ and $\hat{P} \tau_\perp A \tau = 0$, because

$$\begin{aligned}
\tau A \hat{Q} &= \tau A (\tau \hat{Q}) \\
&= \tau A (I - \tau_\perp) \hat{Q} \\
&= \tau A \hat{Q} - \tau A \tau_\perp \hat{Q},
\end{aligned} \tag{60}$$

$$\begin{aligned}
\hat{P} A \tau &= (\hat{P} \tau) A \tau \\
&= \hat{P} (I - \tau_\perp) A \tau \\
&= \hat{P} A \tau - \hat{P} \tau_\perp A \tau.
\end{aligned} \tag{61}$$

(iii) Based on (i) and (ii), $\tau^{(k)} A \tau_\perp^{(k)} \hat{\hat{Q}}^{(k)} \to 0$ and $\hat{\hat{P}}^{(k)} \tau_\perp^{(k)} A \tau^{(k)} \to 0$.
These two quantities should become smaller as the iteration continues.

The numerical convergence can be improved by the following modifications, which can be implemented with t varying from 0 to 1 as the iteration continues, which is referred to as *the homotopy method*.

(i) In Step (2) $A_1^{(k)}$ and $A_2^{(k)}$ are replaced with $t \cdot A_1^{(k)}$ and $t \cdot A_2^{(k)}$, respectively.
(ii) In Step (5) E_{i_r} is replaced with the matrix with 1 at (i_r, i_r) position, $\frac{1-t}{n_m}$ at all the other diagonal entries, and zeros elsewhere.

1.3. MAIN RESULTS

The finite element method (FEM) in conjunction with high–speed digital computers permits an efficient solution to large, complex structural dynamics problems. Many efficient and comprehensive finite element computer codes are now available for structural dynamics response calculations involving harmonic response, transient response, and random response of complex structures. For a certain desired accuracy, however, these codes produce an analytical dynamic model of a large number of degrees of freedom.

Experimental modal analysis, or modal testing, has become an established and popular method for practical structural vibration problems. It utilizes a dedicated modal test equipment, and requires disturbing (by a modal impact hammer or shaker) the structure into motion and recording the resulting motions throughout the structure. Modal testing results in natural frequencies, damping ratios, and mode shapes. These modal data are identified by modal parameter estimation methods, such as circle fit, polyreference method, eigensystem realization algorithm, etc (Allemang and Brown, 1987).

The finite element model tends to have as many an element as possible, in order to describe the structure accurately. Because of this, there are "redundant" nodes from the input–output point of view. On the other hand, the experimental modal model is of low order when compared to the finite element analysis model. Hence, in the work that follows, the finite element analysis model will be reduced to a "smaller" model and be forced to satisfy the orthogonality conditions with the experimental modal data.

The objective in this section is to achieve a reduced model in the state space such that the new state variables are a subset of the original states, which is the new and significant development in the area of model reduction. The method is described in the following.

(1) Analytical Model. Mass (M) and stiffness (K) matrices of order n_1 are derived through finite element analysis. A damping matrix (D) is estimated based on M, K and damping ratios obtained by the experiment (modal testing). The physical model is converted to the state–space form $(A, B, C, \mathbf{x})$ where A is *a system matrix*, B is *an input matrix*, C is *an output matrix*, and $\mathbf{x}$ is *a state vector* composed of generalized velocity and displacement vectors. In the following the system $(A, B, C, \mathbf{x})$ of order n $(n = 2n_1)$ is reduced to a smaller system $(\hat{A}_r, \hat{B}_r, \hat{C}_r, \hat{\mathbf{x}}_r)$ of order n_r $(n_r = 2n_2)$ in a *balanced* coordinate system. In order to preserve the same physical meaning of the state variables, we transform the reduced system $(\hat{A}_r, \hat{B}_r, \hat{C}_r, \hat{\mathbf{x}}_r)$ to $(A_r, B_r, C_r, \mathbf{x}_r)$, keeping $\mathbf{x}_r$ a subset of the original state vector $\mathbf{x}$. The system matrix A_r and input matrix B_r are not compatible with the original state and input matrices due to the compensation for deleted state variables. The system matrix A_r will be further modified by using modal data.

(2) Experimental Model. The m_1 measurement stations are selected to coincide with node points of the finite element model. The structure is disturbed by a modal impact hammer, and its responses from accelerometers are collected by a data–acquisition system. Digitally recorded data are

analyzed, from which extracted are modal parameters such as natural frequencies, damping ratios, and mode shapes. Even though measurements are made at m_1 number of locations, modal parameters identified are usually less than m_1, say m_2 $m_1 \geq m_2$: natural frequencies $\{\omega_i\}_{i=1}^{m_2}$, damping ratios $\{\zeta_i\}_{i=1}^{m_2}$, and mode shape vectors $\{\phi_i \in \mathbf{C}^{m_1}\}_{i=1}^{m_2}$. The mode shape vector $\phi_i \in \mathbf{C}^{m_1}$ is to be expanded to $\phi_i \in \mathbf{C}^{n_1}$ and $\phi_i \in \mathbf{C}^{n_2}$, for them to be used in damping matrix estimation and model modification algorithms. The details are given in Appendix CC.

(3) Reduced Model Modification. The reduced model $(A_r, B_r, C_r, \mathbf{x}_r)$ is in the "same" coordinate system, but A_r and B_r are not compatible with their corresponding original A and B. Therefore, A_r will be modified by the eigenvalues and eigenvectors identified from the modal test results and B_r will be selected directly from B.

The final reduced model $(A_r, B_r, C_r, \mathbf{x}_r)$ is

- $\mathbf{x}_r = \{x_{j_i} | i = 1, \ldots, n_2; \ x_{j_i} \in \mathbf{x}$, where $\mathbf{x} \in \mathbf{R}^n$ is the original state vector}
- $A_r = \{$ Reduced from A and modified to satisfy the eigenvalues and eigenvectors identified from modal testing $\}$
- $B_r = \{j_i^{th} \text{ row of } B | i = 1, \ldots, n_2\}$
- $C_r = \{j_i^{th} \text{ column of } C | i = 1, \ldots, n_2\}$

where $\{j_1, \ldots, j_{n_2}\}$ are indices of those selected state variables.

1.3.1. Model Reduction via Internal Balancing

An analytical model for a linear dynamical system is usually derived through the finite element method. The original equations of motion are written

$$M\ddot{\mathbf{q}}(t) + D\dot{\mathbf{q}}(t) + K\mathbf{q}(t) = B_1\mathbf{u}(t)$$
$$\mathbf{y}(t) = C_1\dot{\mathbf{q}}(t) + C_2\mathbf{q}(t) \tag{62}$$

where

(i) $\mathbf{q}(t) \in \mathbf{R}^{n_1}$, $\mathbf{u}(t) \in \mathbf{R}^p$, and $\mathbf{y}(t) \in \mathbf{R}^r$; $n = 2n_1$;
(ii) M, D, and K are symmetric and positive definite matrices $\in \mathbf{R}^{n_1 \times n_1}$;
(iii) B_1, C_1, and C_2 are real matrices of appropriate dimensions;
(iv) rank$[B_1] = p$, rank$[C_1\ C_2] = r$;
(v) $\mathbf{q}(t) = $ the generalized coordinates; $\mathbf{u}(t) = $ applied forces; $\mathbf{y}(t) = $ outputs.

The system of Equation (64) will be referred to as *a physical* (or *spatial*) *model*. It can be converted to state–space form by the augmented state vector

$$\mathbf{x}(t) = \begin{bmatrix} \mathbf{x}_1(t) \\ \mathbf{x}_2(t) \end{bmatrix} = \begin{bmatrix} \dot{\mathbf{q}}(t) \\ \mathbf{q}(t) \end{bmatrix} \tag{63}$$

Then

$$\begin{bmatrix} \dot{\mathbf{x}}_1(t) \\ \dot{\mathbf{x}}_2(t) \end{bmatrix} = \begin{bmatrix} -M^{-1}D & -M^{-1}K \\ I & 0 \end{bmatrix} \begin{bmatrix} \mathbf{x}_1(t) \\ \mathbf{x}_2(t) \end{bmatrix} + \begin{bmatrix} M^{-1}B_1 \\ 0 \end{bmatrix} \mathbf{u}(t)$$

$$\mathbf{y}(t) = [\,C_1 \quad C_2\,] \begin{bmatrix} \mathbf{x}_1(t) \\ \mathbf{x}_2(t) \end{bmatrix} \tag{64}$$

The system of Equation (64) will be simply written as

$$\dot{\mathbf{x}}(t) = A\mathbf{x}(t) + B\mathbf{u}(t)$$
$$\mathbf{y}(t) = C\mathbf{x}(t) \tag{65}$$

where

$$A = \begin{bmatrix} -M^{-1}D & -M^{-1}K \\ I & 0 \end{bmatrix} \quad B = \begin{bmatrix} -M^{-1}B_1 \\ 0 \end{bmatrix} \tag{66}$$
$$C = [\,C_1 \quad C_2\,]$$

and will be denoted by $(A, B, C, \mathbf{x})$. It is assumed that the system $(A, B, C, \mathbf{x})$ is controllable, observable, and asymptotically stable. Here "asymptotically stable" means that the real parts of the system eigenvalues are all negative.

Let us define *a controllability grammian* (W_c) and *an observability grammian* (W_o) in the following way:

$$W_c = \int_0^\infty e^{At} B B^T e^{A^T t} dt$$
$$W_o = \int_0^\infty e^{A^T t} C^T C e^{At} dt \tag{67}$$

where e^{At} is *the state transition matrix* of the open–loop system $\dot{\mathbf{x}}(t) = A\mathbf{x}(t)$.

The grammian matrices W_c and W_o are the unique symmetric positive definite matrices that satisfy the Lyapunov matrix equations.

$$A W_c + W_c A^T = -B B^T$$
$$A^T W_o + W_o A = -C^T C \tag{68}$$

One can solve for W_c and W_o, using the algorithm given in Bartels and Stewart (1972).

When the order of matrices, W_c and W_o, is small, the Lyapunov matrix equations could be solved as simple algebraic equations. A brief description of conversion into algebraic equations is given in the following.

Let us suppose that we are solving for X in the following equation

$$PXQ + (PXQ)^T = R = R^T \tag{69}$$

where X and R are symmetric. If index notation is used,

$$(PXQ)_{ij} = P_{ik} X_{kl} Q_{lj} \tag{70}$$

where repeating indices imply summation. The Lyapunov matrix equation is written in summation notation as

$$P_{ik} X_{kl} Q_{lj} + Q_{ki} X_{kl} P_{jl} = R_{ij} \tag{71}$$

Factoring out X_{kl} yields the expression

$$(P_{ik} Q_{lj} + P_{jl} Q_{ki}) \cdot X_{kl} = R_{ij} \tag{72}$$

Because of symmetry, $R_{ij} = R_{ji}$ and $X_{kl} = X_{lk}$, the coefficients of X_{kl} are given by

$$\begin{cases} P_{ik} Q_{lj} + P_{jl} Q_{ki}, & \text{if } k = l; \\ (P_{ik} Q_{lj} + P_{jl} Q_{ki}) + (P_{il} Q_{kj} + P_{jk} Q_{li}), & \text{if } k > l. \end{cases} \tag{73}$$

The equivalent algebraic equation can be expressed in terms of the coefficient matrix and the vectors composed of the lower triangular parts of R and X.

$$
\underbrace{\frac{n(n+1)}{2} \times \frac{n(n+1)}{2}}_{\begin{bmatrix} \text{coefficient} \\ \text{matrix} \end{bmatrix}}
\overbrace{\begin{bmatrix} X_{11} \\ \vdots \\ X_{n1} \\ X_{22} \\ \vdots \\ X_{n2} \\ X_{33} \\ \vdots \\ X_{n3} \\ \vdots \\ X_{nn} \end{bmatrix}}^{(k,l)}
=
\overbrace{\begin{bmatrix} R_{11} \\ \vdots \\ R_{n1} \\ R_{22} \\ \vdots \\ R_{n2} \\ R_{33} \\ \vdots \\ R_{n3} \\ \vdots \\ R_{nn} \end{bmatrix}}^{(i,j)}
\tag{74}
$$

where (k, l) indicates the column index of a coefficient matrix; (i, j) the row index. Thus, the $(n \times n)$ Lyapunov matrix equation is converted to $\frac{n(n+1)}{2}$ linear algebraic equations.

The controllability and observability grammians are used to define measures of controllability and observability in certain directions of the state space. Since the grammians are not invariant under coordinate transformations, there exists a coordinate system in which the grammians are equal and diagonal (Moore, 1981). The corresponding system representation is called *balanced*. A reduced order model can be obtained from the balanced representation by deleting the least controllable and, therefore, least observable part of the balanced state matrix.

A coordinate transformation $\mathbf{x} = P\hat{\mathbf{x}}$ yields an equivalent system denoted by

$$
\begin{aligned}
\dot{\hat{\mathbf{x}}}(t) &= \hat{A}\hat{\mathbf{x}}(t) + \hat{B}\mathbf{u}(t) \\
\mathbf{y}(t) &= \hat{C}\hat{\mathbf{x}}(t)
\end{aligned}
\tag{75}
$$

where $\hat{A} = P^{-1}AP$, $\hat{B} = P^{-1}B$, $\hat{C} = CP$. It is important to observe that

$$
\begin{aligned}
e^{\hat{A}t}\hat{B} &= P^{-1}e^{At}B \\
\hat{C}e^{\hat{A}t} &= Ce^{At}P
\end{aligned}
\tag{76}
$$

The grammians, W_c and W_o, are expressed as coordinate–dependent quantities:

$$
\begin{aligned}
W_c(P) &= P^{-1}\left(\int_0^\infty e^{At}BB^T e^{A^T t}dt\right)P^{-T} \\
&= \int_0^\infty e^{\hat{A}t}\hat{B}\hat{B}^T e^{\hat{A}^T t}dt \\
W_o(P) &= P^T\left(\int_0^\infty e^{A^T t}C^T Ce^{At}dt\right)P \\
&= \int_0^\infty e^{\hat{A}^T t}\hat{C}^T \hat{C}e^{\hat{A}t}dt
\end{aligned}
\tag{77}
$$

DEFINITION: *Internally balanced coordinate (or simply Balanced coordinate). There exists a coordinate system in which the two grammians, $W_c(P)$ and $W_o(P)$, are equal and diagonal, i.e.,*

$$
\begin{aligned}
W_c(P) &= W_o(P) = \Sigma \\
\Sigma &= diag\{\sigma_1, \sigma_2, \cdots, \sigma_n\}
\end{aligned}
\tag{78}
$$

Such a coordinate system is called a balanced coordinate.

REMARK: In a balanced coordinate system the state variables can be arranged according to their controllability and observability.

DEFINITION: *Second–order modes. The singular values of H defined by*

$$H = \Lambda_o^{-1/2} V_o^T V_c \Lambda_c^{-1/2} \qquad (79)$$

will be represented by $\{\sigma_1, \sigma_2, \ldots, \sigma_n\}$ with $\sigma_1 \geq \sigma_2 \geq \cdots \geq \sigma_n > 0$ and will be referred to as second–order modes of the system, where Λ_o, V_o, V_c and Λ_c are from the eigenanalysis of W_c and W_o and thus satisfy the relations

$$\begin{aligned} V_c^T W_c V_c &= \Lambda_c \\ V_o^T W_o V_o &= \Lambda_o \end{aligned} \qquad (80)$$

REMARK: The second–order modes are invariant under coordinate transformations and will be used in determining a relative error during a model reduction procedure.

The internally balanced system will be written as follows, denoted by $(\hat{A}, \hat{B}, \hat{C}, \hat{x})$,

$$\begin{aligned} \dot{\hat{x}}(t) &= \hat{A}\hat{x}(t) + \hat{B}u(t) \\ y(t) &= \hat{C}\hat{x}(t) \end{aligned} \qquad (81)$$

where the original system (A, B, C, x) and the balanced $(\hat{A}, \hat{B}, \hat{C}, \hat{x})$ are related in the following way

$$\begin{aligned} \hat{A} &= P^{-1}AP, \\ \hat{B} &= P^{-1}B, \\ \hat{C} &= CP, \\ \hat{x} &= P^{-1}x, \end{aligned} \qquad (82)$$

The two systems are thus related by a non–singular transformation. This is denoted by

$$P : (A, B, C, x) \Longrightarrow (\hat{A}, \hat{B}, \hat{C}, \hat{x}) \qquad (83)$$

$$\det(P) \neq 0 \qquad (84)$$

The matrix P that transforms the original system (A, B, C, x) into a balanced system $(\hat{A}, \hat{B}, \hat{C}, \hat{x})$ can be obtained using one of the following two algorithms:

(I) Method by Moore(1981)

(i) Solve for W_c and find eigenvalues (Λ_c) and eigenvectors (V_c) such that $V_c^T W_c V_c = \Lambda_c$. Then define $P_1 = V_c \Lambda_c^{-1/2}$.

(ii) The coordinate transformation $\mathbf{x} = P_1 \tilde{\mathbf{x}}$ yields an intermediate system $(\tilde{A}, \tilde{B}, \tilde{C}, \tilde{\mathbf{x}})$ calculated by

$$\tilde{A} = P_1^{-1} A P_1$$
$$\tilde{B} = P_1^{-1} B$$
$$\tilde{C} = C P_1$$

(iii) Solve for $\tilde{W}_o$ and find eigenvalues ($\tilde{\Lambda}_o$) and eigenvectors ($\tilde{V}_o$) such that $\tilde{V}_o^T \tilde{W}_o \tilde{V}_o = \tilde{\Lambda}_o$. Let $P_2 = \tilde{V}_o \tilde{\Lambda}_o^{-1/4}$.

(iv) Another coordinate transformation $\hat{\mathbf{x}} = P_2 \tilde{\mathbf{x}}$ yields the desired balanced system $(\hat{A}, \hat{B}, \hat{C}, \hat{\mathbf{x}})$:

$$\hat{A} = P_2^{-1} \tilde{A} P_2 = P_2^{-1}(P_1^{-1} A P_1) P_2$$
$$\hat{B} = P_2^{-1} \tilde{B} = P_2^{-1} P_1^{-1} B$$
$$\hat{C} = \tilde{C} P_2 = C P_1 P_2$$

(v) The transformation P is given by P_1 and P_2 as

$$P = P_1 P_2.$$

(II) Method by Laub(1980)

(i) Solve for W_c and W_o

$$A W_c + W_c A^T = -B B^T$$
$$A^T W_o + W_o A = -C^T C$$

(ii) Calculate the Cholesky decomposition of W_c:

$$W_c = L_c L_c^T$$

where L_c is a lower triangular matrix.

(iii) Calculate the matrix $L_c^T W_o L_c$.

(iv) Solve the symmetric eigenvalue problem defined by

$$U^T (L_c^T W_o L_c) U = \Lambda^2$$

(v) P is then calculated as

$$P = L_c U \Lambda^{-1/2}$$

REMARK: Laub's algorithm is claimed to be faster than Moore's.

The balanced system is now partitioned as

$$
\begin{bmatrix} \dot{\hat{\mathbf{x}}}_r \\ \dot{\hat{\mathbf{x}}}_d \end{bmatrix} = \begin{bmatrix} \hat{A}_r & \hat{A}_{12} \\ \hat{A}_{21} & \hat{A}_{22} \end{bmatrix} \begin{bmatrix} \hat{\mathbf{x}}_r \\ \hat{\mathbf{x}}_d \end{bmatrix} + \begin{bmatrix} \hat{B}_r \\ \hat{B}_d \end{bmatrix} \mathbf{u}
$$
$$
\mathbf{y} = [\,\hat{C}_r \quad \hat{C}_d\,] \begin{bmatrix} \hat{\mathbf{x}}_r \\ \hat{\mathbf{x}}_d \end{bmatrix}
$$

(85)

$$
P : (A, B, C, \mathbf{x}) \Longrightarrow (\hat{A}, \hat{B}, \hat{C}, \hat{\mathbf{x}})
$$

(86)

where $\hat{A}_r \in \mathbf{R}^{(n-k)\times(n-k)}$, and the state variables have been permuted so that $\sigma_1 \geq \sigma_2 \geq \cdots \geq \sigma_n > 0$. The state variables $\{\hat{x}_1, \hat{x}_2, \ldots, \hat{x}_n\}$ are ordered in such a way that $\{\hat{x}_1 =$ the most controllable and observable state $\}, \cdots, \{\hat{x}_n =$ the least controllable and observable state $\}$. The reduced model $(\hat{A}_r, \hat{B}_r, \hat{C}_r, \hat{\mathbf{x}}_r)$ of order $(n - k)$ can be obtained from the balanced representation by deleting k number of the least controllable and observable states.

A relative error due to model reduction is defined by (Moore, 1981)

$$
\text{Relative error} = \frac{\left[\displaystyle\sum_{i=n-k+1}^{n} \sigma_i^2\right]^{1/2}}{\left[\displaystyle\sum_{i=1}^{n-k} \sigma_i^2\right]^{1/2}}
$$

(87)

where the numerator is composed of second–order modes of deleted states and the denominator is composed of those remaining states.

REMARK:

(i) The model reduction is based on the impulse response, which is similar to a hammer test in vibration testing.
(ii) If the original system is asymptotically stable, then the two subsystems $(\hat{A}_r, \hat{B}_r, \hat{C}_r)$ and $(\hat{A}_d, \hat{B}_d, \hat{C}_d)$ are asymptotically stable. (Moore, 1981; Pernebo and Silverman, 1982)
(iii) The relationship between the original system $(A, B, C, \mathbf{x})$ and the reduced system $(\hat{A}_r, \hat{B}_r, \hat{C}_r, \hat{\mathbf{x}}_r)$ can be established through the deleted states $\hat{\mathbf{x}}_d$, which will be investigated extensively in the following section.

EXAMPLE: In order to demonstrate the procedure discussed in this section, we selected this example from Moore (1981). Note that it has no particular physical meaning. The arrays of $(A, B, C, \mathbf{x})$ and u are given as follows:

$$A = \begin{bmatrix} 0 & 0 & 0 & -150 \\ 1 & 0 & 0 & -245 \\ 0 & 1 & 0 & -113 \\ 0 & 0 & 1 & -19 \end{bmatrix} \qquad B = \begin{bmatrix} 4 \\ 1 \\ 0 \\ 0 \end{bmatrix}$$

$$C = [0 \ 0 \ 0 \ 1]$$

$$u = \delta(t) \quad \dots \quad \text{impulse}$$

(88)

The grammians are

$$W_c = \begin{bmatrix} 13.5326 & 6.7944 & 1.0537 & 0.0533 \\ 6.7944 & 3.5646 & 0.5784 & 0.0298 \\ 1.0537 & 0.5784 & 0.0979 & 0.0051 \\ 0.0533 & 0.0298 & 0.0051 & 0.0003 \end{bmatrix}$$

$$W_o = 10^{-5} \times \begin{bmatrix} 1.5394 & 0. & -2.3067 & 0. \\ 0. & 2.3067 & 0. & -29.745 \\ -2.3067 & 0. & 29.745 & 0. \\ 0. & -29.745 & 0. & 3015.1 \end{bmatrix}$$

(89)

Then the transformation matrix $P : (A, B, C, \mathbf{x}) \Longrightarrow (\hat{A}, \hat{B}, \hat{C}, \hat{\mathbf{x}})$ and P^{-1} are

$$P = \begin{bmatrix} 29.090 & -4.056 & 0.553 & -0.310 \\ 14.784 & 5.449 & -0.557 & 0.426 \\ 2.323 & 2.093 & -0.030 & -0.122 \\ 0.118 & 0.131 & 0.056 & 0.007 \end{bmatrix}$$

$$P^{-1} = Q = \begin{bmatrix} 0.025 & 0.019 & 0.001 & -0.052 \\ -0.041 & 0.032 & 0.263 & 0.852 \\ 0.072 & -0.233 & -0.169 & 14.655 \\ -0.244 & 0.971 & -3.631 & 10.080 \end{bmatrix}$$

(90)

Second–order modes are

$$\{\sigma_1, \sigma_2, \sigma_3, \sigma_4\} =$$

$$\{1.5938 \times 10^{-2}, 2.7243 \times 10^{-3}, 1.2720 \times 10^{-4}, 8.0060 \times 10^{-6}\}. \ (91)$$

The internally balanced system matrices are

$$\hat{A} = \begin{bmatrix} -0.4378 & -1.1685 & -0.4143 & -0.0510 \\ 1.1685 & -3.1353 & -2.8352 & -0.3289 \\ -0.4143 & 2.8352 & -12.4753 & -3.2492 \\ 0.0510 & -0.3289 & 3.2492 & -2.9516 \end{bmatrix} \qquad \hat{B} = \begin{bmatrix} 0.1181 \\ -0.1307 \\ 0.0563 \\ -0.0069 \end{bmatrix}$$

$$\hat{C} = [0.1181 \ 0.1307 \ 0.0563 \ 0.0069]$$

(92)

Examination of the singular values suggests that a reduced system $(\hat{A}_r, \hat{B}_r, \hat{C}_r)$ of order three can be obtained by deleting the last row and/or column from $(\hat{A}, \hat{B}, \hat{C})$:

$$\hat{A}_r = \begin{bmatrix} -0.4378 & -1.1685 & -0.4143 \\ 1.1685 & -3.1353 & -2.8352 \\ -0.4143 & 2.8352 & -12.4753 \end{bmatrix} \quad \hat{B}_r = \begin{bmatrix} 0.1181 \\ -0.1307 \\ 0.0563 \end{bmatrix} \quad (93)$$

$$\hat{C}_r = [\, 0.1181 \quad 0.1307 \quad 0.0563 \,]$$

1.3.2. Reduction in the "Original" State Space

The relation between the balanced and original systems, $(\hat{A}, \hat{B}, \hat{C}, \hat{\mathbf{x}})$ and $(A, B, C, \mathbf{x})$, is expressed by

$$P : (A, B, C, \mathbf{x}) \Longrightarrow (\hat{A}, \hat{B}, \hat{C}, \hat{\mathbf{x}}) \tag{94}$$

$$\begin{aligned} \hat{A} &= P^{-1}AP, \\ \hat{B} &= P^{-1}B, \\ \hat{C} &= CP, \\ \hat{\mathbf{x}} &= P^{-1}\mathbf{x}, \end{aligned} \tag{95}$$

The transformation P and its inverse P^{-1} are denoted by

$$\begin{aligned} P &= [p_{ij}] \\ P^{-1} &= Q = [q_{ij}] \end{aligned} \tag{96}$$

The state variables, $\{\hat{\mathbf{x}}_1, \cdots, \hat{\mathbf{x}}_n\}$, are arranged according to the degree of controllability and observability in the following manner:

$$\hat{x}_1 = \sum_j q_{1j} x_j \qquad \text{the most controllable, observable state}$$

$$\vdots$$

$$\hat{x}_i = \sum_j q_{ij} x_j \tag{97}$$

$$\vdots$$

$$\hat{x}_n = \sum_j q_{nj} x_j \qquad \text{the least controllable, observable state}$$

A reduced model of order $(n - k)$ can be obtained from the balanced representation by deleting the k least controllable and observable states. Let $\hat{\mathbf{x}}_r$ be the states retained and $\hat{\mathbf{x}}_d$ be those deleted. Then the balanced system $(\hat{A}, \hat{B}, \hat{C}, \hat{\mathbf{x}})$ can be partitioned as

$$
\begin{bmatrix} \dot{\hat{\mathbf{x}}}_r \\ \dot{\hat{\mathbf{x}}}_d \end{bmatrix} = \begin{bmatrix} \hat{A}_r & \hat{A}_{12} \\ \hat{A}_{21} & \hat{A}_{22} \end{bmatrix} \begin{bmatrix} \hat{\mathbf{x}}_r \\ \hat{\mathbf{x}}_d \end{bmatrix} + \begin{bmatrix} \hat{B}_r \\ \hat{B}_d \end{bmatrix} \mathbf{u}
$$

$$
\mathbf{y} = [\, \hat{C}_r \quad \hat{C}_d \,] \begin{bmatrix} \hat{\mathbf{x}}_r \\ \hat{\mathbf{x}}_d \end{bmatrix}
$$

(98)

Deleting the k least controllable and observable states, i.e., $\hat{\mathbf{x}}_d=0$, yields

$$
\dot{\hat{\mathbf{x}}}_r(t) = \hat{A}_r \hat{\mathbf{x}}_r(t) + \hat{B}_r \mathbf{u}(t)
$$

$$
\mathbf{y}_r(t) = \hat{C}_r \hat{\mathbf{x}}_r(t)
$$

(99)

a reduced model of order $(n - k)$. At the same time forcing $\hat{\mathbf{x}}_d$ to be zero puts k constraints on the original n state variables, $\{x_1, \dots, x_n\}$:

$$
\hat{x}_1 = f(x_1, \dots, x_n)
$$

$$
\vdots
$$

$$
\hat{x}_{n-k} = f(x_1, \dots, x_n)
$$

$$
\hat{x}_{n-(k-1)} = f(x_1, \dots, x_n)
$$

$$
= \sum_{j=1}^{n} q_{(n-k+1)j} x_j = 0 \qquad \text{for } t \in [0, \infty)
$$

(100)

$$
\vdots
$$

$$
\hat{x}_n = f(x_1, \dots, x_n)
$$

$$
= \sum_{j=1}^{n} q_{nj} x_j = 0 \qquad \text{for } t \in [0, \infty)
$$

Thus k states among the original states, $\{x_1, \dots, x_n\}$, can be removed by the k constraints resulting from this "forced dependency". In other words it allows us to choose $(n - k)$ independent state variables from $\{x_1, \dots, x_n\}$. Let those $(n - k)$ selected state variables be denoted by $\{x_{j_1}, \dots, x_{j_{n-k}}\}$, then $\{\hat{x}_1, \dots, \hat{x}_{n-k}\}$ are linear combinations of $\{x_{j_1}, \dots, x_{j_{n-k}}\}$

$$
\hat{x}_1 = f(x_{j_1}, \dots, x_{j_{n-k}})
$$

$$
\vdots
$$

(101)

$$
\hat{x}_{n-k} = f(x_{j_1}, \dots, x_{j_{n-k}})
$$

Therefore, there exists a transformation matrix such that

$$\mathbf{x}_r = P_r \hat{\mathbf{x}}_r \tag{102}$$

where $\hat{\mathbf{x}}_r$ and $\mathbf{x}_r$ are defined by

$$\hat{\mathbf{x}}_r = [\hat{x}_1 \quad \hat{x}_2 \quad \cdots \quad \hat{x}_{n-k}]^T$$
$$\mathbf{x}_r = [x_{j_1} \quad x_{j_2} \quad \cdots \quad x_{j_{n-k}}]^T \tag{103}$$

The reduced order system $(A_r, B_r, C_r, \mathbf{x}_r)$ is

$$\dot{\mathbf{x}}_r(t) = A_r \mathbf{x}_r(t) + B_r \mathbf{u}(t)$$
$$\mathbf{y}_r(t) = C_r \mathbf{x}_r(t) \tag{104}$$

where

$$A_r = P_r \hat{A}_r P_r^{-1}$$
$$B_r = P_r \hat{B}_r \tag{105}$$
$$C_r = P_r^{-1} \hat{C}_r$$

REMARK: $(\hat{A}_r, \hat{B}_r, \hat{C}_r, \hat{\mathbf{x}}_r)$ is the internally balanced system of $(A_r, B_r, C_r, \mathbf{x}_r)$.

There is a systematic method of obtaining P_r from the original transformation matrix P,

$$\mathbf{x} = P \hat{\mathbf{x}} \tag{106}$$

or, in terms of their elements

$$[x_i] = \sum_{j=1}^{n} p_{ij} \hat{x}_j \tag{107}$$

The model reduction, i.e., $\hat{\mathbf{x}}_d = 0$, sets the last k states of the balanced system to zero,

$$\hat{x}_{n-(k-1)} = 0$$
$$\vdots \tag{108}$$
$$\hat{x}_n = 0$$

The x_i's are expressed as linear combinations of $\{\hat{x}_1, \ldots, \hat{x}_{n-k}\}$ by deleting the last k columns of P,

$$
\begin{aligned}
[x_i] &= \sum_{j=1}^{n} p_{ij}\hat{x}_j \\
&= \sum_{j=1}^{n-k} p_{ij}\hat{x}_j \\
&= \begin{bmatrix} p_{11} & \cdots & p_{1(n-k)} \\ \vdots & & \vdots \\ p_{n1} & \cdots & p_{n(n-k)} \end{bmatrix} \begin{bmatrix} \hat{x}_1 \\ \vdots \\ \hat{x}_{n-k} \end{bmatrix}
\end{aligned}
\tag{109}
$$

then by selecting $\{j_1, \ldots, j_{n-k}\}$ rows from the above P matrix the transformation between $\mathbf{x}_r$ and $\hat{\mathbf{x}}_r$ is established

$$
\begin{bmatrix} x_{j_1} \\ \vdots \\ x_{j_{n-k}} \end{bmatrix} = \begin{bmatrix} p_{j_1 1} & \cdots & p_{j_1 n-k} \\ \vdots & & \vdots \\ p_{j_{n-k} 1} & \cdots & p_{j_{n-k} n-k} \end{bmatrix} \begin{bmatrix} \hat{x}_1 \\ \vdots \\ \hat{x}_{n-k} \end{bmatrix}
\tag{110}
$$

Therefore, the transformation matrix P_r of $\mathbf{x}_r = P_r\hat{\mathbf{x}}_r$ is given

$$
P_r = \begin{bmatrix} p_{j_1 1} & \cdots & p_{j_1 n-k} \\ \vdots & & \vdots \\ p_{j_{n-k} 1} & \cdots & p_{j_{n-k} n-k} \end{bmatrix}
\tag{111}
$$

REMARK: The system matrix of reduced order, A_r, can be partitioned as:

$$
\begin{bmatrix} \tilde{A}_D & \tilde{A}_K \\ I(\epsilon) & \mathbf{0}(\epsilon) \end{bmatrix}
\tag{112}
$$

where $I(\epsilon)$ and $\mathbf{0}(\epsilon)$ are defined as

$$
\begin{aligned}
\|I - I(\epsilon)\| &< \epsilon \\
\|\mathbf{0}(\epsilon)\| &< \epsilon \\
\epsilon = \quad &\text{a positive small number}
\end{aligned}
\tag{113}
$$

The identity (I) and zero ($\mathbf{0}$) matrices became $I(\epsilon)$ and $\mathbf{0}(\epsilon)$, respectively, due to the compensation for those deleted states. By the same reason B_r becomes incompatible with the original B. However, C_r remains compatible with the original C. A_r will be further modified in the following section.

EXAMPLE: The procedure discussed in this section is demonstrated through the same system as in the previous example. The system $(A, B, C, \mathbf{x})$ and u are given as follows:

$$A = \begin{bmatrix} 0 & 0 & 0 & -150 \\ 1 & 0 & 0 & -245 \\ 0 & 1 & 0 & -113 \\ 0 & 0 & 1 & -19 \end{bmatrix} \qquad B = \begin{bmatrix} 4 \\ 1 \\ 0 \\ 0 \end{bmatrix} \qquad (114)$$

$$C = [0 \ 0 \ 0 \ 1]$$

$$u = \delta(t) \quad \ldots \quad \text{impulse}$$

Then the transformation matrices $P : (A, B, C, \mathbf{x}) \Longrightarrow (\hat{A}, \hat{B}, \hat{C}, \hat{\mathbf{x}})$ and P^{-1} are

$$P = \begin{bmatrix} 29.090 & -4.056 & .553 & -.310 \\ 14.784 & 5.449 & -.557 & .426 \\ 2.323 & 2.093 & -.030 & -.122 \\ .118 & .131 & .056 & .007 \end{bmatrix}$$

$$(115)$$

$$P^{-1} = Q = \begin{bmatrix} .025 & .019 & .001 & -.052 \\ -.041 & .032 & .263 & .852 \\ .072 & -.233 & -.169 & 14.655 \\ -.244 & .971 & -3.631 & 10.080 \end{bmatrix}$$

Let's suppose that we decide to delete x_3 so that the reduced model contains the three original states, $\{x_1, x_2, x_4\}$. Here we have chosen x_1, x_2 and x_4 as the subset of the original states that we are most interested in.

(i) Delete the least controllable and observable state in the balanced system, i.e., $\hat{x}_4 = 0$. From $\hat{\mathbf{x}} = P^{-1}\mathbf{x}$ this requires that

$$\hat{x}_4 = \sum_{j=1}^{4} q_{4j} x_j$$

$$(116)$$

$$= -.244 x_1 + .971 x_2 - 3.631 x_3 + 10.080 x_4$$

$$= 0$$

This constraint allows us to write the deleted state, x_3, as a linear combination of $\{x_1, x_2, x_4\}$,

$$x_3 = -.067 x_1 + .267 x_2 + 2.777 x_4 \qquad (117)$$

(ii) From $\hat{\mathbf{x}} = P^{-1}\mathbf{x}$:

$$\hat{x}_1 = f(x_1, x_2, x_3, x_4)$$
$$\hat{x}_2 = f(x_1, x_2, x_3, x_4) \qquad (118)$$
$$\hat{x}_3 = f(x_1, x_2, x_3, x_4)$$

and from (i) one constraint among the original states is obtained

$$x_3 = f(x_1, x_2, x_4) \tag{119}$$

Altogether $\{\hat{x}_1, \hat{x}_2, \hat{x}_3\}$ can be expressed as linear combinations of $\{x_1, x_2, x_4\}$

$$\hat{x}_1 = f(x_1, x_2, x_4)$$
$$\hat{x}_2 = f(x_1, x_2, x_4) \tag{120}$$
$$\hat{x}_3 = f(x_1, x_2, x_4)$$

or, in the matrix form

$$\begin{bmatrix} \hat{x}_1 \\ \hat{x}_2 \\ \hat{x}_3 \end{bmatrix} = P_r^{-1} \begin{bmatrix} x_1 \\ x_2 \\ x_4 \end{bmatrix} \tag{121}$$

Therefore, the transformation between the reduced systems is given by

$$\hat{\mathbf{x}}_r = P_r^{-1} \mathbf{x}_r \tag{122}$$

where

$$\hat{\mathbf{x}}_r = [\hat{x}_1 \quad \hat{x}_2 \quad \hat{x}_3]^T$$
$$\mathbf{x}_r = [x_1 \quad x_2 \quad x_4]^T$$
$$P_r^{-1} = \begin{bmatrix} .025 & .020 & -.049 \\ -.058 & .102 & 1.582 \\ .084 & -.278 & 14.180 \end{bmatrix} \tag{123}$$
$$P_r = \begin{bmatrix} 29.090 & -4.056 & .553 \\ 14.784 & 5.449 & -.557 \\ .118 & .131 & .056 \end{bmatrix}$$

The system matrices of reduced order are

$$A_r = P_r \hat{A}_r P_r^{-1} = \begin{bmatrix} .090 & -.290 & -135.898 \\ .876 & .398 & -264.391 \\ -.069 & .274 & -16.537 \end{bmatrix}$$
$$B_r = P_r \hat{B}_r = \begin{bmatrix} 3.998 \\ 1.003 \\ 0. \end{bmatrix} \tag{124}$$
$$C_r = P_r^{-1} \hat{C}_r = [0 \quad 0 \quad 1]$$

(iii) Let us make a comparison between $(A, B, C, \mathbf{x})$ and $(A_r, B_r, C_r, \mathbf{x}_r)$. The original model is

$$\begin{bmatrix} \dot{x}_1 \\ \dot{x}_2 \\ \dot{x}_3 \\ \dot{x}_4 \end{bmatrix} = A \begin{bmatrix} x_1 \\ x_2 \\ x_3 \\ x_4 \end{bmatrix} + Bu \qquad y = C \begin{bmatrix} x_1 \\ x_2 \\ x_3 \\ x_4 \end{bmatrix} \tag{125}$$

The reduced model is

$$\begin{bmatrix} \dot{x}_1 \\ \dot{x}_2 \\ \dot{x}_4 \end{bmatrix} = A_r \begin{bmatrix} x_1 \\ x_2 \\ x_4 \end{bmatrix} + B_r u \qquad y_r = C_r \begin{bmatrix} x_1 \\ x_2 \\ x_4 \end{bmatrix} \qquad (126)$$

1.3.3. Modification of Reduced Model

One of the important and basic relations often used in dynamic modeling is that the measured modes satisfy the theoretical requirement of weighted orthogonality with respect to the mass and stiffness matrices. Such a requirement can only be satisfied by assuming no, or proportional, damping (Berman and Nagy (1983); Berman and Wei (1981); Berman and Flannelly (1971)).

For structures with non–proportional damping the orthogonality conditions are applied *not uniquely* (Fuh, Chen and Berman (1984); Meirovitch (1980), p.213) because of the following reasons: i) the conversion from a physical model into a state space form is not unique, and ii) there are an infinite number of complex modal vectors that are related through linear transformations.

The eigenvalue problem is, however, uniquely defined in the state space form as well as in the physical model. The reduced model in the previous section can then be modified, with minimum change, to satisfy the eigenvalues and eigenvectors that are identified through modal testing and estimation.

The following informaion is required for the procedure of modifying the reduced (analytical) model.

(1) Modal Test Data. It is assumed that a complete modal survey test has been conducted and that the following test data are available:

(a) The complex modal vectors $\{\phi_i \in \mathbf{C}^{m_1} | i = 1, \dots, m_2\}$ measured at m_1 stations where $m_1 > m_2$. $\phi_i \in \mathbf{C}^{m_1}$ are expanded to $\phi_i \in \mathbf{C}^{n_2}$ through the full mode computation (see Appendix C for details).
(b) The natural frequencies $\{\omega_i | i = 1, \dots, m_2\}$.
(c) The damping ratios $\{\zeta_i | i = 1, \dots, m_2\}$.

(2) Analytical Model. The state–space form of the FEM model of order $(2n_1 \times 2n_1)$ is reduced to the order of $(2n_2 \times 2n_2)$. The system matrix of the reduced model is

$$\begin{bmatrix} \tilde{A}_D & \tilde{A}_K \\ I(\epsilon) & \mathbf{0}(\epsilon) \end{bmatrix} \qquad (127)$$

where $I(\epsilon)$ and $0(\epsilon)$ are defined as

$$\|I(\epsilon) - I\| < \epsilon$$
$$\|0(\epsilon)\| < \epsilon \tag{128}$$
$$\epsilon = \text{a positive small number}$$

The eigenvalue problem for a physical model is written as

$$(M\lambda_i^2 + D\lambda_i + K)\phi_i = 0 \tag{129}$$

where $\lambda_i = -\zeta_i\omega_i \pm j\omega_i\sqrt{1 - \zeta_i^2}$. By defining two matrices

$$\Lambda = \text{diag}\{\lambda_1, \ldots, \lambda_{m_2}\} \in \mathbf{C}^{m_2 \times m_2}$$
$$\Phi = [\phi_1 \quad \phi_2 \quad \cdots \quad \phi_{m_2}] \in \mathbf{C}^{n_2 \times m_2} \tag{130}$$

equation (131) can be written as

$$M\Phi\Lambda^2 + D\Phi\Lambda + K\Phi = 0 \tag{131}$$

The eigenvalue problem in the state space form is

$$\begin{bmatrix} \Phi\Lambda \\ \Phi \end{bmatrix} \Lambda = \begin{bmatrix} -M^{-1}D & -M^{-1}K \\ I & 0 \end{bmatrix} \begin{bmatrix} \Phi\Lambda \\ \Phi \end{bmatrix} \tag{132}$$

or,

$$\Phi\Lambda^2 + (M^{-1}D)\Phi\Lambda + (M^{-1}K)\Phi = 0 \tag{133}$$

For brevity, let $A_D = -M^{-1}D$ and $A_K = -M^{-1}K$.

Our objective is to modify Equation (127) to satisfy Equation (132) with minimum change on $\tilde{A}_D$ and $\tilde{A}_K$.

Problem statement:

$$\min_{A_D, A_K} \varepsilon = \|A_D - \tilde{A}_D\|^2 + \|A_K - \tilde{A}_K\|^2$$
$$\text{subject to} \quad \Phi\Lambda^2 + (-A_D)\Phi\Lambda + (-A_K)\Phi = 0. \tag{134}$$

Let

$$\Phi = P + iQ$$
$$\Phi\Lambda = R + iS \tag{135}$$
$$\Phi\Lambda^2 = U + iV$$

where P, Q, R, S, U and $V \in \mathbf{R}^{n_2 \times m_2}$. Then the constraint can be written

$$(U + iV) + (-A_D)(R + iS) + (-A_K)(P + iQ) = 0 \tag{136}$$

The real and imaginary parts of the constraint are separated into two constraints

$$A_D R + A_K P - U = 0$$
$$A_D S + A_K Q - V = 0$$

(137)

The cost function is

$$\varepsilon = \sum_{i=1}^{n_2} \sum_{j=1}^{n_2} \left[(A_{Dij} - \tilde{A}_{Dij})^2 + (A_{Kij} - \tilde{A}_{Kij})^2 \right]$$

(138)

The cost function and the constraints can be combined by introducing two Lagrange multiplier matrices X and $Y \in \mathbf{R}^{n_2 \times m_2}$ by

$$\psi = \varepsilon + X(A_D R + A_K P - U) + Y(A_D S + A_K Q - V)$$

(139)

Necessary conditions for minimizing ψ with respect to Λ_D and Λ_K, that is,

$$\frac{\partial \psi}{\partial A_D} = 0 \quad \text{and} \quad \frac{\partial \psi}{\partial A_K} = 0$$

(140)

yield

$$A_D = \tilde{A}_D - \frac{1}{2}(X R^T + Y S^T)$$
$$A_K = \tilde{A}_K - \frac{1}{2}(X P^T + Y Q^T)$$

(141)

where Lagrange multiplier matrices are

$$X = \left[2(\tilde{A}_D R + \tilde{A}_K P - U) - Y \cdot (S^T R + Q^T P) \right] \cdot \left(R^T R + P^T P \right)^{-1}$$
$$Y = 2 \left[(\tilde{A}_D S + \tilde{A}_K Q - V) - (\tilde{A}_D R + \tilde{A}_K P - U) \cdot Z \right]$$
$$\cdot \left[S^T S + Q^T Q - (S^T R + Q^T P) \cdot Z \right]^{-1}$$

(142)

and

$$Z = \left(R^T R + P^T P \right)^{-1} \left(R^T S + P^T Q \right)$$

(143)

The detailed derivation is given in Appendix A. The above gives a solution to the modification problem for adjusting the finite element model to fit experimental data.

1.3.4. Damping Matrix Estimation

Damping is important and sometimes critical in the evaluation of structural response. Yet, the determination of damping parameters, more generally the actual model of energy dissipation, has not been systematically studied. Although a general methodology such as finite element analysis has been developed for mass and stiffness matrices, the damping is usually restricted to be *proportional*.

The term proportional damping usually refers to the fraction of critical damping or the modal damping ratio for a particular natural mode of structural vibration. It can be determined in various ways by measuring the rate of amplitude decay in free vibration, the bandwidth of the frequency response function near resonance, or the rate at which the phase angle of response changes as the frequency of sinusoidal excitation is swept through resonance.

References for this section are Beliveau (1976); Caughey (1960); Hasselman (1976(a), 1976(b), 1972); Hendrickson and Inman (1985); and Ibrahim (1983(a), 1983(b)).

It is assumed that the following two models are available.

(1) Analytical Model. Mass and stiffness matrices of order $(n_1 \times n_1)$ are derived through the finite element method.

(2) Modal Model. Modal estimation identifies m_2 number of damping ratios $\{\zeta_i | i = 1, \ldots, m_2\}$, natural frequencies $\{\omega_i | i = 1, \ldots, m_2\}$ and mode shape vectors $\{\phi_i \in \mathbf{C}^{m_1} | i = 1, \ldots, m_2\}$ even though measurements are made at m_1 locations $(m_1 > m_2)$. $\phi_i \in \mathbf{C}^{m_1}$ is expanded to $\phi_i \in \mathbf{C}^{n_1}$ through the full mode computation (see Appendix C for details).

The proportional damping matrix D^0 can be obtained by using damping ratios $\{\zeta_i | i = 1, \ldots, m_2\}$ up to the m_2^{th} mode. It is reasonable to assume that the higher analytical modes, from $m_2 + 1$ to n_1, have a damping ratio equal to average damping factor of the m_2 measured modes (Ibrahim, 1983(b)), i.e.,

$$\zeta = \left(\frac{1}{m_2}\right) \sum_{i=1}^{m_2} \zeta_i \tag{144}$$

The objective is to fit modal data to build a full non–proportional damping matrix D, $D = D^0 + X$, by adjusting X to satisfy the eigenvalue problem with minimum error (Hendrickson and Inman, 1985). The eigenvalue problem for a spatial model is

$$(M\lambda_i^2 + D\lambda_i + K)\phi_i = 0 \quad i = 1, \ldots, m_2 \tag{145}$$

where $\lambda_i = -\zeta_i\omega_i \pm j\omega_i\sqrt{1-\zeta_i^2}$. Equation (145) can be written as

$$D\phi_i = \frac{1}{\lambda_i}\left(-\lambda_i^2 M - K\right)\phi_i$$
$$= \mathbf{f}_i \quad \text{for} \quad i = 1,\ldots,m_2 \tag{146}$$

in matrix form

$$D\Phi = F \tag{147}$$

where

$$\Phi = [\phi_1 \quad \cdots \quad \phi_{m_2}] \in \mathbf{C}^{n_1 \times m_2}$$
$$F = [\mathbf{f}_1 \quad \cdots \quad \mathbf{f}_{m_2}] \in \mathbf{C}^{n_1 \times m_2} \tag{148}$$

The problem is stated as

$$\min_X \|(D^0 + X)\Phi - F\|^2$$
$$\text{subject to} \quad X = X^T \tag{149}$$

If the minimization is carried out one column (X_i) at a time, there are n_1 objective functions of the form

$$\psi_i = \sum_{j=1}^{m_2}\left\{\left[(D^0_i + X_i)^T R_j - P_{ij}\right]^2 + \left[(D^0_i + X_i)^T S_j - Q_{ij}\right]^2\right\} \tag{150}$$

where R, S, P and Q are real and imaginary parts of Φ and F:

$$\Phi = R + iS$$
$$F = P + iQ \tag{151}$$

Starting with the first column, each successive minimization has one less variable than the previous one due to symmetry of X.

The unknown X_{ki}'s are expressed as

$$\sum_{k=i}^{n_1}\left(RR^T + SS^T\right)_{pk} X_{ki} = \left[PR^T + QS^T - D^0\left(RR^T + SS^T\right)\right]_{ip}$$
$$- \sum_{k=1}^{i-1}\left(RR^T + SS^T\right)_{pk} X_{ki}$$
$$\text{for} \quad i = 1,\ldots,n_1$$
$$p = i,\ldots,n_1 \tag{152}$$

The analytical solution for i^{th} objective function is given by solving the following linear equation

$$A^{(i)} X^{(i)} = B^{(i)} \tag{153}$$

where

$$X_k^{(i)} = X_{ki} \quad \text{for} \quad k = i, \dots, n_1; \quad [X_{1i} \cdots X_{(i-1)i}] \text{ are known,}$$

$$A_{pk}^{(i)} = \left(RR^T + SS^T \right)_{pk} \quad \text{for} \quad k = i, \dots, n_1$$

$$B_p^{(i)} = \left[PR^T + QS^T - D^0 \left(RR^T + SS^T \right) \right]_{ip} \tag{154}$$

$$- \sum_{k=1}^{i-1} \left(RR^T + SS^T \right)_{pk} X_{ki}$$

$$\text{for} \quad i = 1, \dots, n_1; \quad p = i, \dots, n_1.$$

Note that a single subscript means a column of the matrix; double subscripts an element of the matrix. The detailed derivation is given in Appendix B.

1.3.5. Short Summary

The procedures described in this Section can be outlined as follows:

(1) Analytical Model

- Finite Element Model: $M, K \in \mathbf{R}^{n_1 \times n_1}$
- Damping Matrix Estimation: $D \in \mathbf{R}^{n_1 \times n_1}$
- State Space Form:

$$\begin{bmatrix} -M^{-1}D & -M^{-1}K \\ I & 0 \end{bmatrix}$$

$$(2n_1 \times 2n_1)$$

- Model Reduction:

$$\begin{bmatrix} \tilde{A}_D & \tilde{A}_K \\ I(\epsilon) & 0(\epsilon) \end{bmatrix}$$

$$(2n_2 \times 2n_2)$$

(2) Experimental Model

- Modal Testing & Identification
- Damping Ratios: $\{\zeta_i | i = 1, \dots, m_2\}$

- Natural Frequencies: $\{\omega_i | i = 1, \dots, m_2\}$
- Mode Shapes: $\{\phi_i \in \mathbf{C}^{m_1} | i = 1, \dots, m_2\}$

$$\Phi \in \mathbf{C}^{m_1 \times m_2}$$

- Full Mode Computation:

$$\Phi \in \mathbf{C}^{n_1 \times m_2} \text{ and } \Phi \in \mathbf{C}^{n_2 \times m_2}$$

(3) Model Modification

- Norm Minimization with eigenvalue and eigenvector Constraint:

$$\begin{bmatrix} A_D & A_K \\ I & 0 \end{bmatrix}$$

$$(2n_2 \times 2n_2)$$

Note:

$$n_1 \geq n_2 \geq m_1 \geq m_2$$

- n_1 Analytical (FEM) Coordinates
- n_2 Reduced Analytical Coordinates
- m_1 Measurement Stations
- m_2 Modal Parameters Identified

The finite element model with an estimated non–proportional damping matrix is converted into the state space model. It is then reduced via the internal balancing method and is expressed in terms of a fewer number of state variables than the original. The reduced model is modified by forcing it to satify the eigenvalues and eigenvectors that are identified from modal testing, instead of requiring the weighted orthogonality relationship on the mass and stiffness matrices.

The final reduced model retains a fewer number of state variables. Each state in the reduced model preserves the same physical meaning as in the original model.

1.4. NUMERICAL ALGORITHM

The numerical algorithm is summarized for programming. The original system is denoted by $(A, B, C, \mathbf{x})$ and its balanced system by $(\hat{A}, \hat{B}, \hat{C}, \hat{\mathbf{x}})$.

The sizes of the matrices are

$$A, \hat{A} \in \mathbf{R}^{2n_1 \times 2n_1}$$
$$B, \hat{B} \in \mathbf{R}^{2n_1 \times p}$$
$$C, \hat{C} \in \mathbf{R}^{q \times 2n_1} \quad (155)$$
$$\mathbf{x}, \hat{\mathbf{x}} \in \mathbf{R}^{2n_1}$$

$$P : (A, B, C, \mathbf{x}) \Longrightarrow (\hat{A}, \hat{B}, \hat{C}, \hat{\mathbf{x}}); \quad \mathbf{x} = P\hat{\mathbf{x}} \quad (156)$$

The reduced system is denoted by $(A_r, B_r, C_r, \mathbf{x}_r)$ and its balanced system by $(\hat{A}_r, \hat{B}_r, \hat{C}_r, \hat{\mathbf{x}}_r)$. The sizes of the matrices are

$$A_r, \hat{A}_r \in \mathbf{R}^{2n_2 \times 2n_2}$$
$$B_r, \hat{B}_r \in \mathbf{R}^{2n_2 \times p}$$
$$C_r, \hat{C}_r \in \mathbf{R}^{q \times 2n_2} \quad (157)$$
$$\mathbf{x}_r, \hat{\mathbf{x}}_r \in \mathbf{R}^{2n_2}$$

$$P_r : (A_r, B_r, C_r, \mathbf{x}_r) \Longrightarrow (\hat{A}_r, \hat{B}_r, \hat{C}_r, \hat{\mathbf{x}}_r); \quad \mathbf{x}_r = P_r\hat{\mathbf{x}}_r \quad (158)$$

(1) Internal Balancing (Laub, 1980)

(i) Solve for W_c and W_o

$$A W_c + W_c A^T = -B B^T$$
$$A^T W_o + W_o A = -C^T C$$

(ii) Calculate the Cholesky decomposition of W_c,

$$W_c = L_c L_c^T$$

where L_c is a lower triangular matrix
(iii) Calculate the matrix $L_c^T W_o L_c$
(iv) Solve the symmetric eigenvalue/eigenvector problem,

$$U^T (L_c^T W_o L_c) U = \Lambda^2$$

(v) P is then calculated as

$$P = L_c U \Lambda^{-1/2}$$

(vi) The coordinate transformation, $\mathbf{x} = P\hat{\mathbf{x}}$, yields

$$\hat{A} = P^{-1}AP$$
$$\hat{B} = P^{-1}B$$
$$\hat{C} = CP$$

(2) Model Reduction

(i) The reduced model in the balanced coordinate can be found as follows

$$\hat{A}_r = \text{Select first } n_2 \text{ columns and rows from} \hat{A}$$
$$= \left[\hat{A}_{ij} \right] \text{ for } i, j = 1, \ldots, n_2$$
$$\hat{B}_r = \text{Select first } n_2 \text{ rows from} \hat{B}$$
$$= \left[\hat{B}_{ij} \right] \text{ for } i = 1, \ldots, n_2; \ j = 1, \ldots, p$$
$$\hat{C}_r = \text{Select first } n_2 \text{ columns from} \hat{C}$$
$$= \left[\hat{C}_{ij} \right] \text{ for } i = 1, \ldots, q; \ j = 1, \ldots, n_2$$

(ii) Select the state variables to be retained from $\{x_1, \ldots, x_{n_1}\}$. Let the indicies of those selected be $\{j_1, \ldots, j_{n_2}\}$.
(iii) The transformation matrix P_r can be obtained by selecting first n_2 columns and $\{j_1, \ldots, j_{n_2}\}$ rows from P.
(iv) The reduced model in terms of the original state variables is now expressed as

$$A_r = P_r \hat{A}_r P_r^{-1}$$
$$B_r = P_r \hat{B}_r$$
$$C_r = \hat{C}_r P_r^{-1}$$

(3) Modification of A_r

(i) A_r is partitioned as

$$A_r = \begin{bmatrix} \tilde{A}_D & \tilde{A}_K \\ I(\epsilon) & 0(\epsilon) \end{bmatrix}$$

(ii) A_D and A_K are calculated by

$$A_D = \tilde{A}_D - \frac{1}{2}\left(XR^T + YS^T\right)$$

$$A_K = \tilde{A}_K - \frac{1}{2}\left(XP^T + YQ^T\right)$$

where $P, Q, R, S, X,$ and Y are defined in Section 3.3.

(iii) The modified A_r is

$$A_r = \begin{bmatrix} A_D & A_K \\ I & 0 \end{bmatrix}$$

(4) Modification of B_r

The modified B_r is obtained by selecting $\{j_1, \ldots, j_{n_2}\}$ rows from B.

1.5. NUMERICAL EXAMPLE

This section demonstrates the results in Section 3 by using the numerical algorithm given in Section 4. All the calculations are performed by using MATLAB by The MathWorks, Inc.

The physical model is the one with the simplest geometry, a cantilever beam. Its physical parameters and dimensions are as follows (Pong, 1986):

Length $= 1.835$ m
Thickness $= 0.005918$ m
Width $= 0.10668$ m
Flexual Modulus $(E) = 28$ GPa
Mass Density $= 1624.8$ kg/m^3

The beam is equally divided into nine elements so that it has nine active node points. The node points are named as node 1 through node 9 in such a manner that node 1 is at the closest to the base and node 9 at the tip of the beam. Each node point has one degree of freedom, that is, translational displacement. Rotational displacement is considered to be insignificant and therefore is not included.

The equations of motion are written

$$M\ddot{q} + D\dot{q} + Kq = B_1 u$$
$$y = C_1\dot{q} + C_2 q$$

The mass matrix is approximated as lumped masses:

$$M = \begin{bmatrix} .2092 & 0 & 0 & 0 & 0 & 0 & 0 & 0 & 0 \\ 0 & .2092 & 0 & 0 & 0 & 0 & 0 & 0 & 0 \\ 0 & 0 & .2092 & 0 & 0 & 0 & 0 & 0 & 0 \\ 0 & 0 & 0 & .2092 & 0 & 0 & 0 & 0 & 0 \\ 0 & 0 & 0 & 0 & .2092 & 0 & 0 & 0 & 0 \\ 0 & 0 & 0 & 0 & 0 & .2092 & 0 & 0 & 0 \\ 0 & 0 & 0 & 0 & 0 & 0 & .2092 & 0 & 0 \\ 0 & 0 & 0 & 0 & 0 & 0 & 0 & .2092 & 0 \\ 0 & 0 & 0 & 0 & 0 & 0 & 0 & 0 & .1046 \end{bmatrix}$$

The stiffness matrix is obtained as follows:

$$K = \begin{bmatrix}
1.1462 & -.7248 & .2921 & -.0783 & .0210 & -.0056 & .0015 \\
-.7248 & .8933 & -.6570 & .2739 & -.0734 & .0197 & -.0052 \\
.2921 & -.6570 & .8751 & -.6522 & .2726 & -.0730 & .0195 \\
-.0783 & .2739 & -.6522 & .8738 & -.6518 & .2724 & -.0726 \\
.0210 & -.0734 & .2726 & -.6518 & .8736 & -.6514 & .2711 \\
-.0056 & .0197 & -.0730 & .2724 & -.6514 & .8723 & -.6465 \\
.0015 & -.0052 & .0195 & -.0726 & .2711 & -.6465 & .8542 \\
-.0004 & .0013 & -.0049 & .0182 & -.0678 & .2529 & -.5788 \\
.0001 & -.0002 & .0008 & -.0030 & .0113 & -.0422 & .1573
\end{bmatrix}$$

$$\begin{bmatrix}
-.0004 & .0001 \\
.0013 & -.0002 \\
-.0049 & .0008 \\
.0182 & -.0030 \\
-.0678 & .0113 \\
.2529 & -.0422 \\
-.5788 & .1573 \\
.6012 & -.2219 \\
-.2219 & .0979
\end{bmatrix} \times 10^5$$

The damping ratios are assumed to be 0.002 at every mode. Then the proportional damping matrix is calculated as

$$D = \begin{bmatrix}
.5733 & -.2303 & .0424 & -.0067 & .0023 & -.0002 & .0002 \\
-.2303 & .4355 & -.2338 & .0386 & -.0065 & .0022 & -.0000 \\
.0424 & -.2338 & .4228 & -.2371 & .0372 & -.0069 & .0023 \\
-.0067 & .0386 & -.2371 & .4200 & -.2385 & .0364 & -.0070 \\
.0023 & -.0065 & .0372 & -.2385 & .4186 & -.2396 & .0358 \\
-.0002 & .0022 & -.0069 & .0364 & -.2396 & .4169 & -.2409 \\
.0002 & -.0000 & .0023 & -.0070 & .0358 & -.2409 & -.4098 \\
.0001 & .0005 & .0005 & .0030 & -.0054 & .0356 & -.2358 \\
.0001 & .0003 & .0008 & .0011 & .0037 & .0009 & .0380
\end{bmatrix}$$

$$\begin{bmatrix}
.0001 & .0001 \\
.0005 & .0003 \\
.0005 & .0008 \\
.0030 & .0011 \\
-.0054 & .0037 \\
.0356 & .0009 \\
-.2358 & .0380 \\
.3321 & -.1303 \\
-.1303 & .0846
\end{bmatrix}$$

It should be noted that the numerical algorithm in Section 4 can handle a non–proportional damping matrix as well. By defining the state vector

$$\mathbf{x}(t) = \begin{bmatrix} \mathbf{x}_1(t) \\ \mathbf{x}_2(t) \end{bmatrix} = \begin{bmatrix} \dot{\mathbf{q}}(t) \\ \mathbf{q}(t) \end{bmatrix}$$

the system matrices are written as

$$A = \begin{bmatrix} -M^{-1}D & -M^{-1}K \\ I & 0 \end{bmatrix} \qquad B = \begin{bmatrix} -M^{-1}B_1 \\ 0 \end{bmatrix}$$
$$C = [\, C_1 \;\; C_2 \,]$$

For the purpose of demonstration the impulse input is placed on node 5 (mid point) and the displacement of node 9 (the tip) is measured:

$$B_1 = [0 \;\; 0 \;\; 0 \;\; 0 \;\; 1 \;\; 0 \;\; 0 \;\; 0 \;\; 0]^T$$
$$C_1 = 0$$
$$C_2 = [0 \;\; 0 \;\; 0 \;\; 0 \;\; 0 \;\; 0 \;\; 0 \;\; 0 \;\; 1]$$

Now let us assume that nodes 3, 5, 7, and 9 are selected along with their displacement and velocity, so that the selected states are j_i = $\{3, 5, 7, 9, 12, 14, 16, 18\}$. The reduced–order model $(A_r, B_r, C_r, \mathbf{x}_r)$ is

$$A_r = \begin{bmatrix}
-.3932 & .2405 & -.1624 & .0658 \\
.0352 & -.3929 & .4729 & -.1689 \\
.3010 & .0213 & -.4813 & .2395 \\
-.4930 & .2589 & .4592 & -.3528 \\
1.0000 & 2.8118 \times 10^{-5} & -1.5991 \times 10^{-6} & -3.9513 \times 10^{-6} \\
5.0654 \times 10^{-5} & 1.0000 & 2.5257 \times 10^{-6} & 6.3348 \times 10^{-6} \\
-3.1052 \times 10^{-5} & 2.7687 \times 10^{-5} & 1.0000 & -3.7552 \times 10^{-6} \\
2.2895 \times 10^{-5} & -2.0956 \times 10^{-5} & 1.1181 \times 10^{-6} & 1.0000
\end{bmatrix}$$

$$\begin{bmatrix}
-1.5395 \times 10^4 & 1.2492 \times 10^4 & -4.2030 \times 10^3 & 3.9175 \times 10^2 \\
1.7118 \times 10^4 & -2.6156 \times 10^4 & 1.8880 \times 10^4 & -5.3672 \times 10^3 \\
-9.0584 \times 10^3 & 2.1852 \times 10^4 & -2.2973 \times 10^4 & 8.5616 \times 10^3 \\
1.0171 \times 10^4 & -2.6641 \times 10^4 & 3.1394 \times 10^4 & -1.2751 \times 10^4 \\
-1.6115 \times 10^{-3} & 1.1701 \times 10^{-2} & -1.5558 \times 10^{-2} & 6.3328 \times 10^{-3} \\
1.2521 \times 10^{-3} & -8.3276 \times 10^{-3} & 1.0962 \times 10^{-2} & -4.4479 \times 10^{-3} \\
-1.4224 \times 10^{-3} & 1.0529 \times 10^{-2} & -1.4028 \times 10^{-2} & 5.7136 \times 10^{-3} \\
9.0739 \times 10^{-3} & -8.1847 \times 10^{-2} & 1.1110 \times 10^{-1} & -4.5518 \times 10^{-2}
\end{bmatrix}$$

$$B_r = \begin{bmatrix} 0.2866 \\ 2.1950 \\ 0.0880 \\ 0.1337 \\ 0. \\ 0. \\ 0. \\ 0. \end{bmatrix}$$

$$C_r = [\,0. \quad 0. \quad 0. \quad 0. \quad 0. \quad 0. \quad 0. \quad 1.0\,]$$

The second–order modes of the system are

$$S = \begin{bmatrix} 1.8964 \\ 1.8888 \\ 8.2180 \times 10^{-2} \\ 8.1852 \times 10^{-2} \\ 3.5102 \times 10^{-3} \\ 3.4962 \times 10^{-3} \\ 2.2903 \times 10^{-3} \\ 2.2812 \times 10^{-3} \\ 4.6111 \times 10^{-4} \\ 4.5927 \times 10^{-4} \\ 2.3207 \times 10^{-4} \\ 2.3115 \times 10^{-4} \\ 8.8535 \times 10^{-5} \\ 8.8183 \times 10^{-5} \\ 2.4828 \times 10^{-5} \\ 2.4729 \times 10^{-5} \\ 1.5594 \times 10^{-5} \\ 1.5532 \times 10^{-5} \end{bmatrix}$$

The relative error associated with model reduction is

$$\text{Relative error} = 2.7635 \times 10^{-4}$$

The modal data from the experiment is simulated, in this example, by taking the four eigenvalues and corresponding mode shape vectors from the original system $(A, B, C, \mathbf{x})$. The eigenvalues are

$$\Lambda = \begin{bmatrix} -.487 + 243.5i & 0 & 0 & 0 \\ 0 & -.252 + 125.9i & 0 & 0 \\ 0 & 0 & -.091 + 45.52i & 0 \\ 0 & 0 & 0 & -.015 + 7.505i \end{bmatrix}$$

The corresponding mode shape vectors are

$$
\Phi = \begin{bmatrix}
3.9405 \times 10^{-4} + 1.9702 \times 10^{-1}i & -1.1124 \times 10^{-3} - 5.5618 \times 10^{-1}i \\
-1.0007 \times 10^{-3} - 5.0034 \times 10^{-1}i & 3.5358 \times 10^{-4} + 1.7679 \times 10^{-1}i \\
1.0172 \times 10^{-3} + 5.0859 \times 10^{-1}i & 7.9976 \times 10^{-4} + 3.9988 \times 10^{-1}i \\
-1.3449 \times 10^{-3} - 6.7243 \times 10^{-1}i & -1.4135 \times 10^{-3} - 7.0676 \times 10^{-1}i
\end{bmatrix}
$$

$$
\begin{bmatrix}
8.8419 \times 10^{-4} + 4.4209 \times 10^{-1}i & -2.5562 \times 10^{-4} - 1.2781 \times 10^{-1}i \\
1.0129 \times 10^{-3} + 5.0647 \times 10^{-1}i & -6.2675 \times 10^{-4} - 3.1337 \times 10^{-1}i \\
7.0011 \times 10^{-5} + 3.5005 \times 10^{-2}i & -1.0740 \times 10^{-3} - 5.3699 \times 10^{-1}i \\
-1.4789 \times 10^{-3} - 7.3947 \times 10^{-1}i & -1.5455 \times 10^{-3} - 7.7272 \times 10^{-1}i
\end{bmatrix}
$$

The new transformation matrix P_r is obtained by selecting the first 8 columns and 3rd, 5th, 7th, 9th, 12th, 14th, 16th, and 18th rows from the original transformation matrix P.

$$
P_r = \begin{bmatrix}
.2948 & -.2969 & 3.3211 & 3.3301 \\
.7234 & -.7275 & 3.8002 & 3.8195 \\
1.2406 & -1.2457 & .2628 & .2638 \\
1.7861 & -1.7916 & -5.5514 & -5.5738 \\
3.9472 \times 10^{-2} & 3.9377 \times 10^{-2} & 7.3035 \times 10^{-2} & -7.3082 \times 10^{-2} \\
9.6727 \times 10^{-2} & 9.6603 \times 10^{-2} & 8.3806 \times 10^{-2} & -8.3587 \times 10^{-2} \\
.1657 & .1656 & 5.9030 \times 10^{-3} & -5.6673 \times 10^{-3} \\
.2383 & .2385 & -.1221 & .1224
\end{bmatrix}
$$

$$
\begin{bmatrix}
4.1005 & -4.2199 & -3.2809 & 3.4511 \\
-1.2021 & 1.4444 & 8.4797 & -8.6193 \\
-3.1006 & 2.8788 & -8.6963 & 8.6861 \\
5.4308 & -5.1381 & 11.533 & -11.449 \\
3.3376 \times 10^{-2} & 3.2711 \times 10^{-2} & -1.4383 \times 10^{-2} & -1.3275 \times 10^{-2} \\
-1.1001 \times 10^{-2} & -9.9911 \times 10^{-3} & 3.5458 \times 10^{-2} & 3.4761 \times 10^{-2} \\
-2.3213 \times 10^{-2} & -2.4318 \times 10^{-2} & -3.5407 \times 10^{-2} & -3.5957 \times 10^{-2} \\
4.1120 \times 10^{-2} & 4.2878 \times 10^{-2} & 4.6521 \times 10^{-2} & 4.7825 \times 10^{-2}
\end{bmatrix}
$$

Then the modified system matrix A_r^{modified} is

$$
A_r^{\text{modified}} = \begin{bmatrix}
-.4357 & .2098 & -.0135 & -.0086 \\
.2684 & -.5091 & .2875 & -.0499 \\
-.0588 & .3079 & -.4316 & .1639 \\
.0691 & -.2907 & .5608 & -.3133 \\
1. & 0. & 0. & 0. \\
0. & 1. & 0. & 0. \\
0. & 0. & 1. & 0. \\
0. & 0. & 0. & 1.
\end{bmatrix}
$$

$$\begin{bmatrix} -1.5394 \times 10^4 & 1.2491 \times 10^4 & -4.2024 \times 10^3 & 3.9159 \times 10^2 \\ 1.7117 \times 10^4 & -2.6154 \times 10^4 & 1.8879 \times 10^4 & -5.3670 \times 10^3 \\ -9.0581 \times 10^3 & 2.1852 \times 10^4 & -2.2973 \times 10^4 & 8.5615 \times 10^3 \\ 1.0171 \times 10^4 & -2.6640 \times 10^4 & 3.1394 \times 10^4 & -1.2751 \times 10^4 \\ 0. & 0. & 0. & 0. \\ 0. & 0. & 0. & 0. \\ 0. & 0. & 0. & 0. \\ 0. & 0. & 0. & 0. \end{bmatrix}$$

$$\underline{B}_r^{\text{modified}} = \begin{bmatrix} 0 \\ 4.7813 \\ 0 \\ 0 \\ 0 \\ 0 \\ 0 \\ 0 \end{bmatrix}$$

The C_r^{modified} is exactly same as C_r.

1.6. CONCLUSION

1.6.1. Summary

The finite element method (FEM) has been used extensively in analyzing dynamic structures analytically, but generates an analytical model of large order for a certain desired accuracy. On the other hand, modal testing and estimation techniques identify an experimental (modal) model of low order. These two models are combined to obtain a smaller analytical model by reducing the analytical model and then modifying it with minimum change to satisfy the experimental modal data.

The analytical model with the estimated non–proportional damping matrix is converted into the state space model. Then it is reduced via the internal balancing method and is expressed in terms of a fewer number of state variables than the original. The reduced model is modified by forcing it to satisfy the eigenvalues and eigenvectors that are identified from modal testing, instead of requiring the (weighted) orthogonality relationship on the mass and stiffness matrices.

The final reduced model retains a less number of state variables. Each state in the reduced model preserves the same physical meaning as in the original model. It reflects not only the original FEM model but also the experimental modal data. Its order is smaller than the original FEM model, which is also more agreeable to work with from the control theorists' point of view.

1.6.2. Continuing Research

There are a number of topics for continuing research based on this work, for example:

- The reduced–order modeling can be performed successively by deleting a few states at a time and modifying it until it reaches to the desired order. This could substantially reduce the error associated with the model reduction.
- As is pointed out in Appendix C, the non–proportional damping matrix estimation and the full mode computation can be carried out iteratively to improve accuracies in both sides. The convergence of iterations is to be investigated.
- The reduced model can be used in designing a reduced–order state estimator.
- The reduced model can be used to derive a feedback control law and eventually control the original system. For that purpose further correction/modification should be done on the system and/or input matrices. The performance degradation should first be studied as Aoki (1968) mentioned.
- It might be feasible to identify reduced–order parameter matrices such as the mass, damping and stiffness matrices from the reduced model.

6.A. APPENDIX A

The cost function is

$$
\begin{aligned}
\psi &= \varepsilon + X(A_D R + A_K P - U) + Y(A_D S + A_K Q - V) \\
&= \sum_{i=1}^{n_2}\sum_{j=1}^{n_2}\left[(A_{Dij} - \tilde{A}_{Dij})^2 + (A_{Kij} - \tilde{A}_{Kij})^2\right] \\
&\quad + \sum_{i=1}^{n_2}\sum_{j=1}^{m_2}\left[X_{ij}(A_D R + A_K P - U)_{ij} + Y_{ij}(A_D S + A_K Q - V)_{ij}\right]
\end{aligned}
$$

(159)

Minimization of ψ with respect to A_D and A_K

$$
\frac{\partial \psi}{\partial A_{Dpq}} = 0 \quad \text{and} \quad \frac{\partial \psi}{\partial A_{Kpq}} = 0 \quad \text{for} \quad p, q = 1, \ldots, n_2; \quad (160)
$$

$$\frac{\partial \psi}{\partial A_{Dpq}} = \sum_{i=1}^{n_2} \sum_{j=1}^{n_2} 2(A_{Dij} - \tilde{A}_{Dij}) \frac{\partial A_{Dij}}{\partial A_{Dpq}}$$

$$+ \sum_{i=1}^{n_2} \sum_{j=1}^{m_2} \left[X_{ij} \left(\sum_{k=1}^{n_2} \frac{\partial A_{Dik}}{\partial A_{Dpq}} R_{kj} \right) + Y_{ij} \left(\sum_{k=1}^{n_2} \frac{\partial A_{Dik}}{\partial A_{Dpq}} S_{kj} \right) \right]$$

$$= \sum_{i=1}^{n_2} \sum_{j=1}^{n_2} 2(A_{Dij} - \tilde{A}_{Dij}) \delta_{ip} \delta_{jq}$$

$$+ \sum_{i=1}^{n_2} \sum_{j=1}^{m_2} \left[X_{ij} \left(\delta_{ip} \sum_{k=1}^{n_2} \delta_{kq} R_{kj} \right) + Y_{ij} \left(\delta_{ip} \sum_{k=1}^{n_2} \delta_{kq} S_{kj} \right) \right]$$

$$= 2(A_{Dpq} - \tilde{A}_{Dpq}) + \sum_{j=1}^{m_2} (X_{pj} R_{qj} + Y_{pj} S_{qj})$$

$$= 2(A_D - \tilde{A}_D)_{pq} + (X R^T + Y S^T)_{pq}$$

$$(161)$$

where δ_{ij} is a substitution operator, i.e.,

$$\begin{cases} \delta_{ij} = 0, & \text{if } i \neq j; \\ \delta_{ij} = 1, & \text{if } i = j. \end{cases} \tag{162}$$

Therefore,

$$\frac{\partial \psi}{\partial A_{Dpq}} = 2(A_D - \tilde{A}_D)_{pq} + (X R^T + Y S^T)_{pq}$$
$$= 0 \quad \text{for} \quad p, q = 1, \dots, n_2 \tag{163}$$

In other word,

$$2(A_D - \tilde{A}_D) + (X R^T + Y S^T) = 0 \tag{164}$$

Then

$$A_D = \tilde{A}_D - \frac{1}{2}(X R^T + Y S^T) \tag{165}$$

Similarly for A_K,

$$A_K = \tilde{A}_K - \frac{1}{2}(X P^T + Y Q^T) \tag{166}$$

$\frac{\partial \psi}{\partial X_{pq}} = 0$ and $\frac{\partial \psi}{\partial Y_{pq}} = 0$ produce two constraints.

$$A_D R + A_K P - U = 0 \tag{167}$$

$$A_D S + A_K Q - V = 0 \tag{168}$$

To solve for Lagrange multiplier matrices X and Y substitute Equations (165) and (166) into (167) and (168).

$$X = \left[2(\tilde{A}_D R + \tilde{A}_K P - U) - Y \cdot (S^T R + Q^T P)\right] \cdot \left(R^T R + P^T P\right)^{-1}$$

$$Y = 2\left[(\tilde{A}_D S + \tilde{A}_K Q - V) - (\tilde{A}_D R + \tilde{A}_K P - U) \cdot Z\right]$$

$$\cdot \left[S^T S + Q^T Q - (S^T R + Q^T P) \cdot Z\right]^{-1}$$

$$(169)$$

where

$$Z = \left(R^T R + P^T P\right)^{-1} \left(R^T S + P^T Q\right) \tag{170}$$

6.B. APPENDIX B

The cost function is

$$\psi_i = \sum_{j=1}^{m_2} \left\{ \left[\left(D^0{}_i + X_i\right)^T R_j - P_{ij}\right]^2 + \left[\left(D^0{}_i + X_i\right)^T S_j - Q_{ij}\right]^2 \right\}$$

subject to $X_{ij} = X_{ji}$

$$(171)$$

Minimization of ψ with respect to X_i

$$\frac{\partial \psi_i}{\partial X_{pi}} = 0 \quad \text{for} \quad p = i, \dots, n_1 \tag{172}$$

utilizing the known quantities $\{X_{1i}, \dots, X_{(i-1)i}\}$ from the $(i-1)$ previous minimizations.

$$\frac{\partial \psi_i}{\partial X_{pi}} \cdot \left(\frac{1}{2}\right) = \sum_{j=1}^{m_2} \left[\left(D^0{}_i + X_i\right)^T R_j - P_{ij}\right] \cdot \left(\frac{\partial}{\partial X_{pi}} \sum_{k=1}^{n_1} X_{ki} R_{kj}\right)$$

$$+ \sum_{j=1}^{m_2} \left[\left(D^0{}_i + X_i\right)^T S_j - Q_{ij}\right] \cdot \left(\frac{\partial}{\partial X_{pi}} \sum_{k=1}^{n_1} X_{ki} S_{kj}\right)$$

$$= \sum_{j=1}^{m_2} \left[\left(D^0{}_i + X_i\right)^T R_j - P_{ij}\right] \cdot \left(\sum_{k=1}^{n_1} \delta_{kp} R_{kj}\right)$$

$$+ \sum_{j=1}^{m_2} \left[\left(D^0{}_i + X_i\right)^T S_j - Q_{ij}\right] \cdot \left(\sum_{k=1}^{n_1} \delta_{kp} S_{kj}\right) \tag{173}$$

$$= \sum_{j=1}^{m_2} \left[\left(D^0{}_i + X_i \right)^T R_j - P_{ij} \right] \cdot R_{pj}$$

$$+ \sum_{j=1}^{m_2} \left[\left(D^0{}_i + X_i \right)^T S_j - Q_{ij} \right] \cdot S_{pj}$$

Since $\{X_{ki} | k = 1, \ldots, i-1\}$ are already known,

$$\frac{\partial \psi_i}{\partial X_{pi}} \cdot \left(\frac{1}{2} \right) = \sum_{j=1}^{m_2} \left[D^{0T}_i R_j - P_{ij} + \sum_{k=1}^{i-1} X_{ki} R_{kj} + \sum_{k=i}^{n_1} X_{ki} R_{kj} \right] \cdot R_{pj}$$

$$+ \sum_{j=1}^{m_2} \left[D^{0T}_i S_j - Q_{ij} + \sum_{k=1}^{i-1} X_{ki} S_{kj} + \sum_{k=i}^{n_1} X_{ki} S_{kj} \right] \cdot R_{pj}$$

$$= \sum_{j=1}^{m_2} \sum_{k=i}^{n_1} X_{ki} \left(R_{kj} R_{pj} + S_{kj} S_{pj} \right) + \sum_{j=1}^{m_2} \sum_{k=1}^{i-1} X_{ki} \left(R_{kj} R_{pj} + S_{kj} S_{pj} \right)$$

$$+ \sum_{j=1}^{m_2} \left[\left(\sum_{k=1}^{n_1} D^0{}_{ki} R_{kj} - P_{ij} \right) R_{pj} + \left(\sum_{k=1}^{n_1} D^0{}_{ki} S_{kj} - Q_{ij} \right) S_{pj} \right]$$

$$= \sum_{k=i}^{n_1} X_{ki} \left(RR^T + SS^T \right)_{kp}$$

$$+ \sum_{k=1}^{i-1} X_{ki} \left(RR^T + SS^T \right)_{kp}$$

$$+ \left[D^0 \left(RR^T + SS^T \right) - \left(PR^T + QS^T \right) \right]_{ip}$$

(174)

Then unknown X_{ki}'s are expressed as

$$\sum_{k=i}^{n_1} \left(RR^T + SS^T \right)_{pk} X_{ki} = \left[PR^T + QS^T - D^0 \left(RR^T + SS^T \right) \right]_{ip}$$

$$- \sum_{k=1}^{i-1} \left(RR^T + SS^T \right)_{pk} X_{ki}$$

$$\text{for} \quad i = 1, \ldots, n_1$$

$$p = i, \ldots, n_1$$

(175)

By defining

$$X_k^{(i)} = X_{ki} \quad \text{for} \quad k = i, \ldots, n_1$$

$$A_{pk}^{(i)} = \left(RR^T + SS^T\right)_{pk} \quad \text{for} \quad k = i, \ldots, n_1$$

$$B_p^{(i)} = \left[PR^T + QS^T - D^0\left(RR^T + SS^T\right)\right]_{ip}$$

$$\qquad\qquad - \sum_{k=1}^{i-1}\left(RR^T + SS^T\right)_{pk} X_{ki}$$

$$\text{for} \quad i = 1, \ldots, n_1$$

$$p = i, \ldots, n_1$$

(176)

$\{X_{ii} X_{(i+1)i} \cdots X_{n_1 i}\}$ can be obtained by solving the linear algebraic equation

$$A^{(i)} X^{(i)} = B^{(i)}. \tag{177}$$

6.C. APPENDIX C

When *a priori* knowledge of mode shape vectors is available, the expansion of $\phi_i \in \mathbf{C}^{m_1}$ into $\phi_i \in \mathbf{C}^{n_1}$ could be a simple curve fitting such as interpolation and /or extrapolation.

The following method is suggested in Fuh, Chen and Berman (1984) for general cases. The mode shape vector $\phi_i \in \mathbf{C}^{n_1}$, which is i^{th} column of the mode shape matrix $\Phi \in \mathbf{C}^{n_1 \times m_2}$, is permuted and divided into two parts: the measured and identified portion, $\phi_i^{(1)} \in \mathbf{C}^{m_1}$, and the unknown portion, $\phi_i^{(2)} \in \mathbf{C}^{(n_1 - m_2)}$,

$$\phi_i = \begin{bmatrix} \phi_i^{(1)} \\ \phi_i^{(2)} \end{bmatrix} \tag{178}$$

The mass, damping and stiffness matrices are permuted in the same sequence. Then the eigenvalue problem is written with the partitioned M, D, and K matrices:

$$\left(\lambda_i^2 \begin{bmatrix} M_1 & M_2 \\ M_2^T & M_4 \end{bmatrix} + \lambda_i \begin{bmatrix} D_1 & D_2 \\ D_2^T & D_4 \end{bmatrix} + \begin{bmatrix} K_1 & K_2 \\ K_2^T & K_4 \end{bmatrix}\right) \cdot \begin{bmatrix} \phi_i^{(1)} \\ \phi_i^{(2)} \end{bmatrix} = 0 \tag{179}$$

The unknown portion can be expressed in term of known quantities

$$\phi_i^{(2)} = -\left(\lambda_i^2 M_4 + \lambda_i D_4 + K_4\right)^{-1} \cdot \left(\lambda_i^2 M_2^T + \lambda_i D_2^T + K_2^T\right) \phi_i^{(1)} \tag{180}$$

REMARK:

(i) The similar technique can be seen in the Guyan's reduction in different context (Guyan, 1965).

(ii) The numerical applicability to a large order system is discussed in Berman and Nagy (1983).

(iii) In order to apply this method the damping matrix must be guessed first, which is usually a proportional damping, D^0. The damping matrix estimation (Section 3.4) and full mode computation can be carried out iteratively to improve accuracies in both sides, but the convergence of iterations is to be investigated.

References

1. Allemang, R.J. and Brown, D.L., 1987, "Modal analysis: twenty years back – twenty years ahead," *Sound and Vibration*, January, pp. 10-16.

2. Aoki, M., 1968, "Control of large–scale dynamics systems by aggregation," *IEEE Trans. Automat. Contr.*, **AC–13**, 246–253.

3. Beliveau, J.-G., 1976, "Identification of viscous damping in structures from modal information," *ASME Trans. J. Appl. Mech.*, **43**, 335–339.

4. Bellman, R., 1960, *Introduction to Matrix Analysis.* McGraw–Hill Book Co.

5. Berman, A. and Flannelly, W.G., 1971, "Theory of incomplete models of dynamic structures," *AIAA Journal*, **9**, 1481–1487.

6. Berman, A. and Nagy, E.J., 1983, "Improvement of a large analytical model using test data," *AIAA Journal*, **21**, 1168–1173.

7. Berman A. and Wei, F.S., 1981, "Automated dynamic analytical model improvement," *NASA Contractor Report 3452.*

8. Bishop, R.E.D. and Gladwell, G.M.L., 1963, "An investigation into the theory of resonance testing," *Phil. Trans. of the Royal Society*, **A255**, 241–280.

9. Caughey, T.K., 1960, "Classical normal modes in damped linear systems," *ASME Trans. J. Appl. Mech.*, **27**(ser.E), 269–271.

10. Caughey, T.K. and O'Kelley, M.E.J., 1965, "Classical normal modes in damped linear dynamic systems," *ASME Trans. J. Appl. Mech.*, **32**, 583–588.

11. Chen, C.-T., 1970, *Introduction to Linear System Theory.* Holt, Rinehart and Winston Inc.

12. Chen, J.C. and Garba, J.A., 1980, "Analytical model improvement using modal test result," *AIAA Journal*, **18**, 684–690.

13. Davison, E.J., 1966, "A method for simplifying linear dynamic systems," *IEEE Trans. Automat. Contr.*, **AC–11**, 93–101.

14. Ewins, D.J., 1986, *Modal Testing: Theory and Practice.* Research Studies Press Ltd., England.

15. Fernando, K.V. and Nicholson, H., 1983, "On the structure of balanced and other principal representations of SISO systems," *IEEE Trans. Automat. Contr.*, **AC–28**, 228–233.

16. Fuh, J.-S. and Berman, A., 1986, "Comment on 'Stiffness matrix adjustment using mode data,' " *AIAA Journal*, **24**, 1405–1406.

17. Fuh, J.-S. and Chen, S.-Y., 1986, "Constraints of the structural modal synthesis," *AIAA Journal*, **24**, 1045–1047.

18. Fuh, J.-S., Chen, S.-Y. and Berman, A., 1984, "System identification of analytical models of damped structures," *25th Structures, Structural Dynamics and Materials Conference*, pp. 112-116.

19. Garwronski, W. and Natke, H.G., 1987(a), "Realizations of the transfer–function matrix," *Int. J. Systems Sci.*, **18**, 229–236.
20. Garwronski, W. and Natke, H.G., 1987(b), "Balancing linear systems," *Int. J. Systems Sci.*, **18**, 237–249.
21. Genesio, R. and Milanese, M., 1976, "A note on the derivation and use of reduced–order models," *IEEE Trans. Automat. Contr.*, **AC–21**, 118–122.
22. Guyan, R.J., 1965, "Reduction of stiffness and mass matrices," *AIAA Journal*, **3**, 380.
23. Hasselman, T.K., 1976(a), "Modal coupling in lightly damped structures," *AIAA Journal*, **14**, 1627–1628.
24. Hasselman, T.K., 1976(b), "Damping synthesis from substructure tests," *AIAA Journal*, **14**, 1409–1418.
25. Hasselman, T.K., 1972, "Method for constructing a full modal damping matrix from experimental measurements," *AIAA Journal*, **10**, 526–527.
26. Hendrickson, W.L. and Inman, D.J., 1985, "Identification of a damping matrix from modal data," Proc. 5th VPI & SU Symposium on Dynamics and Control of Large Structures.
27. Hickin, J. and Sinha, N., 1980, "Model reduction for linear multivariable systems," *IEEE Trans. Automat. Contr.*, **AC–25**, 1121–1127.
28. Hyland, D.C. and Bernstein, D.S., 1985, "The optimal projection equations for model reduction and the relationships among the methods of Wilson, Skelton, and Moore," *IEEE Trans. Automat. Contr.*, **AC–30**, 1201–1211.
29. Ibrahim, S.R., 1983(a), "Dynamic modeling of structures from measured complex modes," *AIAA Journal*, **21**, 898-901.
30. Ibrahim, S.R., 1983(b), "Computation of normal modes from identified complex modes," *AIAA Journal*, **21**, 446–451.
31. Irons, B., 1965, "Structural eigenvalue problem: elimination of unwanted variables," *AIAA Journal*, **3**, 961–962.
32. Kabamba, P.T., 1985, "Balanced gains and their significance for L^2 model reduction," *IEEE Trans. Automat. Contr.*, **AC–30**, 690–693.
33. Kennedy, C.C. and Pancu, C.D.P., 1947, "Use of vectors in vibration measurement and analysis," *J. of the Aeronautical Sci.*, **14**, 603–625.
34. Kidder, R.L., 1973, "Reduction of structural frequency equations," *AIAA Journal*, **11**, 892.
35. Laub, A.J., 1980, "Computation of balancing transformation," *Proc. 1980 JACC*, San Francisco, CA, session FA8–E.
36. Lu, W.–S. and Lee, E.B., 1985, "Model reduction via a quasi–Kalman decomposition," *IEEE Trans. Automat. Contr.*, **AC–30**, 786–790
37. McGrew, J., 1969, "Orthogonalization of measured modes and calculation of influence coefficients," *AIAA Journal*, **7**, 774–776.
38. Meirovitch, L., 1980, *Computational Methods in Structural Dynamics.* Sijthoff and Noordhoff, Rockville, MD.
39. Meirovitch, L., 1967, *Analytical Methods in Vibrations.* The MacMillan Co.
40. Moore, B.C., 1981, "Principal component analysis in linear systems: controllability, observability, and model reduction," *IEEE Trans. Automat. Contr.*, **AC–26**, 17–32.
41. O'Callahan, J., Avitable, P. and Leung, R., 1984, "Development of mass and stiffness matrices for an analytical model using experimental modal data," *Proc. 2nd Int. Modal Analysis Conf.*, pp. 585–592.
42. Pernebo, L. and Silverman, L.M., 1982, "Model reduction via balanced state space representation," *IEEE Trans. Automat. Contr.*, **AC–27**, 382–387.
43. Pong, M.–F., 1986, *Physical Model Improvement by Using Modal Testing Data.* Master's Project, State University of New York at Buffalo.
44. Ramsden, J.N. and Stoker, J.R., 1969, "Mass condensation: a semi–automatic method for reducing the size of vibration problems," *Int. J. for Numr. Meth. in Eng.*, **1**, 333–349.
45. Rieger, N.F., 1986, "The relationship between finite element analysis and modal analysis," *Sound and Vibration*, pp. 16–31.
46. Rodden, W.P., 1967, "A method for deriving structural influence coefficients from ground vibration tests," *AIAA Journal*, **5**, 991–1000.

47. Shames, I.H. and Dym, C.L., 1985, *Energy and Finite Element Methods in Structural Mechanics*. Hemisphere Publishing Corp.
48. Shokoohi, S., Silverman, L.M., and Van Dooren, P.M., 1983, "Linear Time–Variable Systems: Balancing and Model Reduction," *IEEE Trans. Automat. Contr.*, **AC-28**, 810–822.
49. Sinha, N.K. and De Bruin, H., 1973, "Near–optimal control of high–order systems using low–order models," *Int. J. Control*, **17**, 257–262.
50. Skelton, R.E. and Hughes, P.C., 1980, "Modal cost analysis for linear matrix–second–order systems," *ASME Trans. J. Dynamic Sys. Meas. Contr.*, **102**, 151–158.
51. Skelton R.E., Hughes, P.C. and Hablani, H., 1982, "Order reduction for models of space structures using modal cost analysis," *J. Guidance & Control*, **5**, 351–357.
52. Strang, G., 1980, *Linear Algebra and its Applications*. 2nd ed. Academic Press, NY.
53. Thomson, W.T., 1981, *Theory of Vibration with Applications*. Prentice–Hall Inc., NJ.
54. Tse, E.C.Y., Medanic, J.V., and Perkins, W.R., 1978, "Generalized Hessenberg transformations for reduced– order modelling of large–scale systems," *Int. J. Control*, **27**, 493–512.
55. Vaccaro, R.J., 1985, "Deterministic balancing and stochastic model reduction," *IEEE Trans. Automat. Contr.*, **AC–30**, 921–923.
56. Viswanathan, C.N., Longman, R.W., and Likins, P.W., 1984, "A degree of controllability definition: fundamental concepts and application to modal systems," *J. Guidance & Control*, **7**, 222–230.
57. Wilson, D.A., 1974, "Model reduction for multivariable systems," *Int. J. Control*, **20**, 57–64.
58. Wilson, D.A., 1970, "Optimum solution of model–reduction problem," *Proc. IEE*, **117**, 1161–1165.
59. Wilson, D.A. and Mishra, R.N., 1979, "Design of low order estimator using reduced models," *Int. J. Control*, **29**, 447–456.

2 TECHNIQUES IN ACTIVE DYNAMIC STRUCTURAL CONTROL TO OPTIMIZE STRUCTURAL CONTROLLERS AND STRUCTURAL PARAMETERS

MARK J. SCHULZ[1] and DANIEL J. INMAN[2]

[1]*North Carolina A&T State University, Greensboro, NC 27411, USA*
[2]*Virginia Polytechnic Institute and State University, Blacksburg, VA 24061, USA*

2.1. INTRODUCTION

There are various techniques available to optimize either structural parameters[1-4], or structural controllers[5-8], but there are not many techniques that can simultaneously optimize both the structural parameters and the controller.[1,5,9-11] The advantage of integrating the structural and controller optimization problems is that structure and controller interaction is taken into account in the design process and a more efficient overall design (lower control force/lighter weight) can be achieved, and also multi-disciplinary design optimization can be performed. The down side is that the combined optimization problem is more difficult to formulate and solve, and computations are increased. This chapter presents a brief review of current techniques that can optimize both structural and controller parameters and then some new techniques and examples for simultaneously optimizing structures and controllers.

Integrated control design using a linear-quadratic regulator (LQR) or linear-quadratic Gaussian (LQG) control is discussed in[6,*p.*55]. A two-level optimization is used in which the control system is reoptimized for each structural design. The advantage of this approach is that the gains can be found by a standard Riccati equation solver. The disadvantage is that the reduced order model and observer introduce approximation errors into the solution.

A Lagrangian functional is minimized in[7,p.159] to design the structure and optimal controller simultaneously. In this approach, large order sets of nonlinear equations must be solved by numerical integration, and a conjugate gradient search is used to minimize the functional. The technique is computationally intensive, but nonlinear structures and controllers can also be designed.

In[12], an active control technique is presented to assign complex eigenstructure to the closed loop system, and also to optimize the structural controller. This approach is useful to design a constant gain output feedback controller using the minimum number of measurements. The eigenstructure to be assigned can be defined, for example, to decouple the equations of motion in physical coordinates so that the closed-loop mode shapes are orthogonal to the forcing vector. This is a possible approach to design a vibration isolation system for alignment critical applications, or to improve the flying qualities of aircraft. Eigenstructure optimization is used in[13] to design a controller to suppress vibration for a small system. The eigenstructure optimization techniques are restricted to small problems due to the large number of design variables, and optimizing the eigenstructure is difficult because the optimizer can converge to sub-optimal solutions.

A recent consideration in the design of active structures is stability and safety if damage were to occur to the system. One approach to ensure stability is to use "structural feedback" or symmetric positive definite output feedback with collocated actuators and sensors.[5,p.353] In this method the controller may be designed using a reduced-order or an approximate model, or there may be damage to the original structure, but the controller cannot destabilize the system assuming perfect actuators/sensors. A homotopy technique can be used to optimize the control gains and structural parameters and will often converge to the global minimum from a good starting point. Piezoceramic patches and inertial actuators are convenient to use in designing a damage-tolerant system because the actuator and sensor can be easily collocated. In[14], direct model referenced adaptive control is used to maintain control during failure of a second order plant. Performance of a plant that has undergone damage is more difficult to predict than stabililty, and is presently evaluated using simulation.

Further references on techniques for designing linear and nonlinear structures and controllers are given in[15−18] for the design of an active wing[19], for LQR design in the frequency domain[20−25], for efficient structural modeling and optimization procedures[26−27], for design of adaptive structures and controllers[28], for feedforward control,[29-31] for control of nonlinear and chaotic vibrations, and in[32] for active vibration suppression for rotorcraft.

Three recently developed techniques for structure-controller optimization are presented in the following sections. The first two are reverse design techniques based in the frequency domain and optimize selected parameters of the structure/controller to assign pre-specified frequency response functions to the system. The third technique designs nonlinear or linear structures and controllers using the time response.

Traditionally first order state space equations and a reduced basis solution are used to formulate the structural and controller optimization problem. In contrast, the techniques below use the full size system equations in second order form. This takes advantage of the sparsity of structural matrices to reduce computations and obtain a more accurate solution. This approach is becoming more feasible due to recent improvements in computational techniques for large scale systems. In particular, parallel-vector FEM software, Automatic Differentiation for Design Sensitivity Analysis, Optimization software for vector/parallel computations, and Sparse Matrix Solvers are providing the computational power necessary to solve the full second-order models.

2.2. STRUCTURE AND CONTROLLER OPTIMIZATION USING FREQUENCY RESPONSE FUNCTIONS

A Frequency Response Function Optimization (FRFO) method is presented to optimize structural and controller parameters to obtain a desired frequency response at critical degrees of freedom (DOF) of a structure. This technique[33-35] is used when frequency response information is the most convenient approach to tune a structural model. Only the pre-specified Frequency Response Functions (FRFs) at critical points on the structure and through a critical bandwidth are required.

The frequency domain representation is useful because the FRF defines the system's natural frequencies, damping, and modal participation and completely characterizes the steady state harmonic response of the system. This has several advantages over the time domain representation of the response: the response can be decomposed into different frequency bands to focus on the frequency range that is most important in the analysis and to improve signal-to-noise ratios when using test data; the attainable dynamic range (in amplitude) in the frequency domain is usually larger than in the time domain; and the dynamic characteristics of the structure are more clearly shown by plotting the FRF instead of a time response. For active control, the controller can be designed using FRFO by adding feedback matrices to the system equation. Derivation of the method is presented below.

2.2.1. Equations of Motion

The equations of motion that describe an n dimensional second-order linear dynamic system subject to a harmonic input are

$$\mathbf{M\ddot{x}} + \mathbf{D\dot{x}} + \mathbf{Kx} = \text{Real}[\mathbf{p}_o \exp(i\omega t)] \tag{2.1}$$

where $\mathbf{M}$, $\mathbf{D}$, and $\mathbf{K}$ are the mass, damping, and stiffness matrices of the closed-loop system, $\mathbf{x}$ is an $n \times 1$ vector of displacements, t is time, $i = \sqrt{-1}$, and ω is the driving frequency. The bold notation is used to represent matrices and vectors. The system matrices are assumed to be symmetric and positive definite, and $\mathbf{D}$ and $\mathbf{K}$ represent both the structural matrices and the gains for feedback control. Often $\mathbf{M}$ is diagonal, or can be made diagonal. The forcing vector $\mathbf{p}_o$ is constant and possibly complex, and defines the loading points on the structure. The solution to (2.1) is obtained as

$$\mathbf{x}(t) = \text{Real}[\mathbf{q}_o \exp(i\omega t)] \tag{2.2}$$

where $\mathbf{q}_o$ is a complex vector. Substituting (2.2) into (2.1) gives

$$\text{Real}[((\mathbf{K} - \omega^2\mathbf{M} + i\omega\mathbf{D})\mathbf{q}_o - \mathbf{p}_o) \exp(i\omega t)] = \mathbf{O} \tag{2.3}$$

A solution to (2.3) is

$$(\mathbf{K} - \omega^2\mathbf{M} + i\omega\mathbf{D})\mathbf{q}_o = \mathbf{p}_o \tag{2.4}$$

Now partition $\mathbf{q_0}$ as $\mathbf{q_0} = [\mathbf{q_1 q_2}]^T$ where the subscript 1 denotes coordinates where the frequency response is to be assigned, and the subscript 2 denotes coordinates where the frequency response is not critical, and hence is not assigned. Define the inverse of the system receptance matrix as $\mathbf{H}^{-1} = (\mathbf{K} - \omega^2\mathbf{M} + i\omega\mathbf{D})$, which in partitioned form is

$$
\begin{aligned}
\mathbf{H}^{-1} &= \begin{bmatrix} \mathbf{H}_{11} & \mathbf{H}_{12} \\ \mathbf{H}_{21} & \mathbf{H}_{22} \end{bmatrix} \\
&= \begin{bmatrix} (\mathbf{K}_{11} - \omega^2\mathbf{M}_{11} + i\omega\mathbf{D}_{11}) & (\mathbf{K}_{12} - \omega^2\mathbf{M}_{22} + i\omega\mathbf{D}_{22}) \\ (\mathbf{K}_{21} - \omega^2\mathbf{M}_{21} + i\omega\mathbf{D}_{21}) & (\mathbf{K}_{22} - \omega^2\mathbf{M}_{22} + i\omega\mathbf{D}_{22}) \end{bmatrix}
\end{aligned} \tag{2.5}
$$

Rewriting (2.4) gives

$$\begin{bmatrix} \mathbf{H}_{11} & \mathbf{H}_{12} \\ \mathbf{H}_{21} & \mathbf{H}_{22} \end{bmatrix} \begin{bmatrix} \mathbf{q}_1 \\ \mathbf{q}_2 \end{bmatrix} = \begin{bmatrix} \mathbf{p}_1 \\ \mathbf{p}_2 \end{bmatrix} \tag{2.6}$$

Eliminating the $\mathbf{q}_2$ coordinates from equation (2.6) gives

$$(\mathbf{H}_{11} - \mathbf{H}_{12}\mathbf{H}_{22}^{-1}\mathbf{H}_{21})\mathbf{q}_1 - (\mathbf{p}_1 - \mathbf{H}_{12}\mathbf{H}_{22}^{-1}\mathbf{p}_2) = 0 \tag{2.7}$$

Inverting $\mathbf{H}_{22}$ in (2.7) as a function of frequency for each iteration of an optimization is computationally very expensive for large order systems. This limitation can be lessened by restricting the **D** matrix to have the same connectivity and symmetry as the **K** matrix, although it is not necessarily proportional to **K**. This restriction also makes sense on a physical basis because material-dependent damping has the same connectivity as the stiffness matrix, and because the velocity and position feedback control forces would be applied at the same location on the structure. Then from (2.5)

$$\mathbf{H}_{22}(\omega) = (\mathbf{K}_{22} - \omega^2 \mathbf{M}_{22} + i\omega \mathbf{D}_{22} \tag{2.8}$$

Equation (2.8) shows that for a typical finite element structural model the $\mathbf{H}_{22}$ matrix is sparse, symmetric, and has the same banding as the **K** matrix. Thus its inverse can be found using a sparse matrix solver such as in MATLAB.[36] A further reduction in the computations can be obtained by assuming that the structure has proportional damping on the elemental level. This eliminates the structural damping design variables from the optimization.

Computational requirements of the technique can also be reduced by carefully building the finite element model to minimize the bandwidth and number of non-zero entries in the stiffness matrix. This is done by minimizing the number of elements in the model and the number of contiguous elements per node, and using repetitive geometry and symmetry, and, if possible, using lower order elements (e.g., use plate or beam elements instead of solid elements). The next step in the procedure is to define the objective function, which is shown below.

2.2.2. Objective Function

An objective function that can be minimized to "assign" the $\mathbf{q}_1(\omega)$ frequency response vector to the system is written using (2.8) as

$$J = \sum_{r=1}^{nf} \mathbf{E}_r^{*T} \mathbf{Q} \mathbf{E}_r \tag{2.9}$$

where **Q** is a diagonal weighting matrix, nf is the number of frequency points used to define the frequency response curves, T is transpose, $*$ is complex conjugate, and $\mathbf{E}_r$ is the error term which is written below.

$$\mathbf{E}_r = [(\mathbf{H}_{11} - \mathbf{H}_{12}\mathbf{H}_{22}^{-1}\mathbf{H}_{21})\mathbf{q}_1 - (\mathbf{p}_1 - \mathbf{H}_{12}\mathbf{H}_{22}^{-1}\mathbf{p}_2)]_{\omega=\omega_r} \tag{2.10}$$

This form of error function was found to be much simpler (smoother) to optimize than an error function defined as the least squares difference between the actual and assigned FRFs. Moreover, the cost function J can be frequency shaped by making **Q** depend on frequency ω to focus on critical frequency ranges in the optimization. In many cases the $\mathbf{p}_2$ vector in (2.10) will contain all zeros, and this will simplify computations.

There are two possible levels of parameter optimization that can be used with the technique. The first level, presented here, updates direct entries of the structural matrices. The second level optimizes primary design parameters such as material or section properties, and can also perform a feature based parametric optimization. The complexity of the model and computational considerations drive the level of parametization. Repetition in the entries of the system matrices can be taken advantage of by "linking design variables"[2,37] to greatly reduce the number of optimization variables. Selecting the design variables as direct entries of the updating matrices takes advantage of the sparsity of structural matrices, retains matrix connectivity and symmetry of the model, and preserves the physical significance of the model. Additionally, geometric characteristics of the model can be preserved, and redesign of only certain sections of the model can be specified.

The design variables are are taken as the nonzero unique entries of the upper triangular parts of the system matrices, $\mathbf{M}$, $\mathbf{D}$ and $\mathbf{K}$, and are automatically selected by specifying the coordinate DOF for which the system matrices are to be adjusted. Once a design vector is set up, the design variables will then be automatically linked by a searching algorithm to maintain identical properties for "families of elements" and to retain symmetry, connectivity, and geometric similarity in the updated model. A standard optimization routine, such as contained within the Matlab optimization toolbox[38], can be used to minimize (2.9). The only constraints on the design variables will be bounds on their magnitudes to ensure that sign changes or unreasonable magnitudes of matrix elements do not occur. The design variable bounds are given as follows.

$$\xi_j^L \leq \xi_j \leq \xi_j^U \qquad j = 1, 2 \ldots, ndv \tag{2.11}$$

where ξ_j^L and ξ_j^U represent the lower and upper bounds on the design variables ξ_j, and ndv is the total number of design variables in the problem. No functional constraints are needed with this formulation. In order to make the optimization computationally feasible, a closed form gradient of the objective function is derived. As shown below, the gradient can be obtained exactly without any additional function evaluations or matrix inversions. Since the second order form of the system equations is used, matrix sparsity is preserved and the MATLAB sparse matrix functions take advantage of this to further streamline the computational process.

2.2.3. Gradient of the Objective Function

The closed form gradient of the objective function (2.9) is:

$$\frac{\partial J}{\partial \xi_j} = \sum_{r=1}^{nf} 2\text{Real}(s_r) \tag{2.12}$$

where

$$s_r = \mathbf{E}_r^{*T} \mathbf{Q} \frac{\partial \mathbf{E}_r}{\partial \xi_j} \quad \text{and} \quad \frac{\partial \mathbf{E}^{*T}}{\partial \xi_j} = \left(\frac{\partial \mathbf{E}_r}{\partial \xi_j} \right)^{*T}$$

The term $\partial \mathbf{E}_r / \partial \xi_j$ only involves matrix addition and multiplication to compute, and has the value

$$\frac{\partial \mathbf{E}_r}{\partial \xi_j} = \left[\left(\frac{\partial \mathbf{H}_{11}}{\partial \xi_j} - \frac{\partial \mathbf{H}_{12}}{\partial \xi_j} \mathbf{H}_{22}^{-1} \mathbf{H}_{21} - \mathbf{H}_{12} \left(\frac{\partial \mathbf{H}_{22}^{-1}}{\partial \xi_j} \mathbf{H}_{21} + \mathbf{H}_{22}^{-1} \frac{\partial \mathbf{H}_{21}}{\partial \xi_j} \right) \right) \mathbf{q}_q$$
$$+ \left(\frac{\partial \mathbf{H}_{12}}{\partial \xi_j} \mathbf{H}_{22}^{-1} + \mathbf{H}_{12} \frac{\partial \mathbf{H}_{22}^{-1}}{\partial \xi_j} \right) \mathbf{p}_2 \right]_{\omega=\omega_2} \qquad (2.14)$$

where

$$\frac{\partial \mathbf{H}_{\alpha\beta}}{\partial \xi_j} = \frac{\partial (\mathbf{K}_{\alpha\beta} - \omega^2 \mathbf{M}_{\alpha\beta} + i\omega \mathbf{D}_{\alpha\beta}}{\partial \xi_j},$$
$$\alpha, \beta = 1 \text{ or } 2 \text{ and } \frac{\partial \mathbf{H}_{22}^{-1}}{\partial \xi_j} = -\mathbf{H}_{22}^{-1} \frac{\partial \mathbf{H}_{22}}{\partial \xi_j} \mathbf{H}_{22}^{-1}.$$

The gradients of the system matrices $\mathbf{M}, \mathbf{D}$ and $\mathbf{K}$, with respect to the design variables are simply matrices with all zeros, except with ones in the locations corresponding to the locations of the particular design variable. In programming the technique, the gradient matrices can be generated using the sparse matrix function in MATLAB. Since both the objective function and gradient calculations require looping over the nf assigned frequencies, they can be computed together at each frequency point to save computer time and storage. The gradient calculation also loops on the design variables at each frequency point.

The computer algorithms to test the method were developed using the Matlab software system and were run on a 90 Mhz PC computer. The optimization step is performed using the CONSTR subroutine contained in the Matlab optimization toolbox. The algorithm that CONSTR uses to minimize the objective function is described in[38]. The CONSTR algorithm is a non-linear optimization routine that finds the constrained minimum of a scalar multi-variable function. The optimization is performed by arranging the design variables in a reduced design vector containing the structural variables that have been linked together. The only constraints used in this technique are upper and lower bounds on each design variable.

When selecting the DOFs to assign frequency response data, it is helpful to look for the columns of the stiffness matrix with the least number of zeros, that is, the greatest connectivity. Assigning the FRFs to the DOFs with the most connectivity will provide the greatest freedom in tuning the model and will also place the maximum number of zero entries in the H_{22} matrix, which will speed-up computations. Design performance requirements also dictate selection of the DOFs to be assigned.

2.2.4. Optimization of a Truss Structure

This example demonstrates how the FRFO method can be used for the design of a 16 DOF truss structure to minimize vibration. A desired frequency response is assigned to DOF 14 (node 7, y axis) of the four bay truss structure shown in Figure 1. The design goal is to reduce the bending vibration of the truss. A sinusoidal force also acts at DOF 14, as shown in Figure 1. The system M and K matrices are

$$M = \text{diag}[5.6\ 5.6\ 5.6\ 5.6\ 5.6\ 5.6\ 5.6\ 5.6\ 5.6\ 5.6\ 5.6\ 5.6\ 5.5\ 5.5\ 5.3\ 5.3]^{*}10^{-5}$$

$$K = \begin{bmatrix}
2354 & 0.354 & 0 & 0 & -1 & 0 & -0.354 & -0.354 & 0 & 0 & 0 & 0 & 0 & 0 & 0 & 0 \\
 & 1.354 & 0 & -1 & 0 & 0 & -0.354 & -0.354 & 0 & 0 & 0 & 0 & 0 & 0 & 0 & 0 \\
 & & 1.354 & 0.354 & 0 & 0 & -1 & 0 & 0 & 0 & 0 & 0 & 0 & 0 & 0 & 0 \\
 & & & 1.354 & 0 & 0 & 0 & 0 & 0 & 0 & 0 & 0 & 0 & 0 & 0 & 0 \\
 & & & & 2.354 & 0.354 & 0 & 0 & -1 & 0 & -0.354 & -0.354 & 0 & 0 & 0 & 0 \\
 & & & & & 1.354 & 0 & -1 & 0 & 0 & -0.354 & -0.354 & 0 & 0 & 0 & 0 \\
 & & & & & & 2.708 & -0.354 & 0.354 & -1 & 0 & 0 & 0 & 0 & 0 & 0 \\
 & & & & & & & 1.708 & 0.354 & -0.354 & 0 & 0 & 0 & 0 & 0 & 0 \\
 & & & & & & & & 2.354 & -0.354 & 0 & 0 & -1 & 0 & 0 & 0 \\
 & & & & & & & & & 1.354 & 0 & -1 & 0 & 0 & 0 & 0 \\
 & & & & & & & & & & 2.708 & 0 & -0.354 & 0.354 & -1 & 0 \\
 & \text{symm} & & & & & & & & & & 1.708 & 0.354 & -0.354 & 0 & 0 \\
 & & & & & & & & & & & & 2.354 & -0.354 & 0 & 0 \\
 & & & & & & & & & & & & & 1.354 & 0 & -1 \\
 & & & & & & & & & & & & & & 1.354 & -0.354 \\
 & & & & & & & & & & & & & & & 1.354
\end{bmatrix}$$

The vector of displacements is $x(t) = [x1\ y1\ x2\ y2 \ldots x8\ y8]^{T}$. The mass matrix contains about 90% non-structural mass at each node of the truss, and proportional damping ($D = 0.0002K$) and $Q = I$ is assumed to simplify the example.

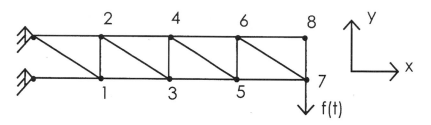

Figure 1. Finite Element Model of a Four Bay Truss.

Design variable linking is used to keep the same form of the stiffness matrix but reduce the number of optimization design variables. Of the forty-four non-zero upper triangular entries in the stiffness matrix, which would be the number of unreduced stiffness design variables, only eight remain after linking repetitive entries. This produces a large computational savings in the solution. Further, note that if the truss were one hundred bays still only eleven design variables would be needed.

The simple linking strategy used in this example does not guarantee that the updated stiffness/gain matrix can be physically realized using tube-type truss members or a particular controller configuration. This technique, however, produces a matrix that identifies the stiffness values that the designer should try to achieve in order for the active structure to have the pre-specified dynamic characteristics. The baseline structural stiffness matrix can be subtracted from the optimized system **K** matrix and the difference becomes the position feedback gain matrix for active control. A scalar optimization factor multiplying each elemental stiffness matrix could also be used to guarantee a physically realizable structural configuration. Shape and area optimization of the truss could also be performed

To perform the analysis, rows and columns 1 and 14 of the system matrices are switched and the repeated inversion of $\mathbf{H}_{22}$ throughout the optimization is performed using the sparse matrix functions in MATLAB. The frequency response of the existing structure at nodes 3 and 7 (DOF 6/14) in the y axis is shown as the solid line in Figure 2. Seven frequency points from 20 to 50 rad/s with a constant value of $-1.7606e - 1 + 1.0144e - 3 * i$ (magnitude $= 0.1761$) are assigned to DOF 14 to flatten the sharp resonance peak shown at 37 rad/s in Figure 2. Stiffness matrix values were arbitrarily bounded so that they would not change more than 50%. The resulting optimization solution approximated the flat FRF in the 20-50rad/s region and shifted the first vibration mode frequency to 80 rad/s. The FRF of the adjusted model is shown as the dashed line in Figure 2.

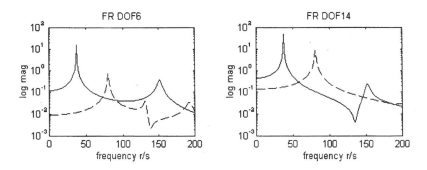

Figure 2. Frequency Response Optimization for Truss Structure (Solid line = original design, dashed line = modified design).

The peak displacement amplitude at the first resonance has been reduced by a factor of 6 by this FRF assignment technique. The adjusted stiffness matrix changed by 47.5% and the eigenvalues changed by 21.4%. The original versus adjusted design vectors are:

$$\mathbf{Dv}(orig) = [790\ 130\ 65\ 725\ 984\ 259\ 363\ 622]^T$$
$$\mathbf{Dv}(adj) = [1128.4\ 150.3\ 32.5\ 1087.5\ 1476.0\ 388.5\ 181.5\ 832.7]^T$$

This example demonstrates that the FRFO technique is a precise method of tailoring the frequency response of a structure while retaining the desired physical significance and connectivity of the dynamic model. More elaborate applications of frequency response assignment can be made, including assigning the phase response relationship between structural components.

2.2.5. Section Summary

The FRFO technique directly assigns frequency response data to simultaneously optimize structural and controller designs. This approach does not require any eigenvalue or eigenvector or basis vector calculations or eigenvalue/vector derivatives, and does not use Guyan or mass reduction, and therefore is an accurate procedure for design optimization.

Limitations of the technique are that it is difficult to define the best FRFs to assign to a model, and parallel computing and automatic differentiation may be required to apply the technique to large models.

2.3. DESIGN OF STRUCTURES AND CONTROLLERS BY FREQUENCY RESPONSE FUNCTION ASSIGNMENT

This section presents a Frequency Response Function Assignment (FRFA) algorithm that assigns a specified Frequency Response Function (FRF) to the structural model. The approach is analogous to control-based Eigenstructure Assignment Theory[13], except that the FRF is assigned rather than complex valued eigenvalues/vectors. The advantage of this formulation is that it is not as computationally intensive as optimizing the FRFs. Assigning a specified FRF to the analytical model of the structure will produce variations in the elemental matrices that represent either structural modifications, or gain coefficients in the design optimization. The FRFA method is a more accurate design procedure than adjusting only structural natural frequencies[39-42] because it uses all the information contained in each FRF, and not just the information around the peaks as in natural frequency adjustment procedures. The FRFA algorithm also includes possible non-proportional damping in the update. This is important as the magnitude of the FRF near resonance is dominated by damping, and the actual structural damping and active damping need to be modeled carefully. The derivation of the technique follows.

2.3.1. Derivation of the FRFA Technique

The equations of motion that describe the linear structure with a harmonic forcing function are

$$\mathbf{M\ddot{x}} + (\mathbf{D} + \overline{\mathbf{D}})\dot{\mathbf{x}} + (\mathbf{K} + \overline{\mathbf{K}})\mathbf{x} = \text{Real}(\mathbf{f}(\omega)\exp(i\omega t)) \qquad (3.1)$$

where $\mathbf{M}$, $\mathbf{D}$ and $\mathbf{K}$ are the mass, damping and stiffness matrices, respectively, of the original or baseline design of the structure, $\overline{D}$ and $\overline{K}$ are the unknown update matrices that represent the redesign modifications to the structure and/or control matrices, $\mathbf{x}(t)$ is an $n \times 1$ vector of displacements, and $\mathbf{f}(\omega)$ is a forcing vector. The matrices $\mathbf{M}$, $\mathbf{D}$ and $\mathbf{K}$ represent a small size finite element model of the structure. The steady state solution to (3.1) is

$$\mathbf{x}(t) = \text{Real}([\mathbf{x}_1^T \ \mathbf{x}_s^T]^T \exp(i\omega t)) \qquad (3.2)$$

where $\mathbf{x}_1$ are $\mathbf{x}_2$ complex vectors representing the degrees of freedom (DOF) where the vibration response is assigned and not assigned, respectively. The solution in (3.2) is used to derive the receptance (displacement/force) frequency response, however, the mobility (velocity/force) frequency response can be derived by taking the derivative of (3.2), that is the velocity response, $\mathbf{v}(t) = i\omega\mathbf{x}(t)$. Since the velocity spectrum is flatter than displacement or acceleration, the mobility function may be numerically more accurate for spectrum comparison and analysis and is used in the following derivation, although any combination of the three quantities can be assigned. Substituting $\mathbf{x}(t) = \mathbf{v}(t)/i\omega$ into (3.1) gives

$$\text{Real}\left[\left([\mathbf{K} + \overline{\mathbf{K}} - \omega^2\mathbf{M} + i\omega(\mathbf{D} + \overline{\mathbf{D}})][\mathbf{v}_1^T\ \mathbf{v}_2^T]^T\frac{1}{i\omega} - \mathbf{f}(\omega)\right)\exp(i\omega t)\right] = 0$$
(3.3)

A solution to (3.3) is

$$[\mathbf{K} + \overline{\mathbf{K}} - \omega^2\mathbf{M} + i\omega(\mathbf{D} + \overline{\mathbf{D}})][\mathbf{v}_1^T\ \mathbf{v}_2^T]^T\frac{1}{i\omega} = \mathbf{f}(\omega) \qquad (3.4)$$

Let $\mathbf{H}^{-1}(\omega) = [\mathbf{K} + \overline{\mathbf{K}} - \omega^2\mathbf{M} + i\omega(\mathbf{D} + \overline{\mathbf{D}}]$ which is the inverse of the receptance frequency response function matrix. Note that it is often practical to assume that the damping matrices have the same connectivity as the stiffness matrices. Then $\mathbf{H}^{-1}$ is a symmetric and very sparse matrix, and computations can be greatly reduced by using sparse matrix functions. In partitioned form (3.4) becomes

$$\begin{bmatrix} \mathbf{H}_{11} & \mathbf{H}_{12} \\ \mathbf{H}_{21} & \mathbf{H}_{22} \end{bmatrix}\begin{bmatrix} \mathbf{v}_1 \\ \mathbf{v}_2 \end{bmatrix}\left(\frac{1}{i\omega}\right) = \begin{bmatrix} \mathbf{f}_1 \\ \mathbf{f}_2 \end{bmatrix} \qquad (3.5)$$

where $\mathbf{H}_{11}, \mathbf{H}_{12} = \mathbf{H}_{21}$, are $\mathbf{H}_{22}$ the partitions of $\mathbf{H}^{-1}$. Since the $\mathbf{v}_2$ coordinates are assumed not measured or assigned, they can be found in terms of the $\mathbf{v}_1$ coordinates by solving the second equation in (3.5) using dynamic expansion as

$$\mathbf{v}_2 = \mathbf{H}_{22}^1(i\omega\mathbf{f}_2 - \mathbf{H}_{21}\mathbf{v}_1) \qquad (3.6)$$

Note that in (3.6), $\mathbf{H}_{22}$ and $\mathbf{H}_{21}$ are not completely known as they contain parts of $\overline{\mathbf{K}}$ and $\overline{\mathbf{D}}$. Thus (3.6) will be exact when the redesign is only contained in $\mathbf{H}_{11}$. To solve for the update matrices $\overline{\mathbf{D}}$ and $\overline{\mathbf{K}}$, we use $\mathbf{v}_2$ from (3.6) and rewrite (3.4) as

$$[\overline{\mathbf{K}}\ \overline{\mathbf{D}}]\begin{bmatrix} \mathbf{I} \\ i\omega_j\mathbf{I} \end{bmatrix}v_j = i\omega_j\mathbf{f}(\omega_j) - [\mathbf{K} + i\omega_j\mathbf{D} - \omega_j^2\mathbf{M}]v_j \qquad (3.7)$$

where $\mathbf{v}_j = [\mathbf{v}_{1j}^T\ \mathbf{v}_{2j}^T]^T$ is the velocity frequency response function vector (v) evaluated at the frequency point ω_j. Equation (3.7) is used as the basis to redesign the structure by using a desired number of frequency points specified at all degrees-of-freedom contained in the $\mathbf{H}_{11}$ partition of the structure. Equation (3.7) represents two sets of equations and should be separated into real and imaginary parts. Let $v_j = v_{Rj} + iv_{Ij}$, where R and I denote the real and imaginary parts. Substituting this into (3.7) gives

$$[\overline{\mathbf{K}}\ \overline{\mathbf{D}}]\begin{bmatrix} \mathbf{v}_{Rj} + i\mathbf{v}_{Ij} \\ -\omega_j\mathbf{v}_{Ij} + i\omega_j\mathbf{v}_{Rj} \end{bmatrix} =$$
$$(-\omega_j\mathbf{f}_{Ij} + (-\mathbf{K} + \omega_j^2\mathbf{M})\mathbf{v}_{Rj} + \omega_j\mathbf{D}\mathbf{v}_{Ij})$$
$$+(\omega_j\mathbf{f}_{Rj} + (-\mathbf{K} + \omega_j^2\mathbf{M})\mathbf{v}_{Ij} - \omega_j\mathbf{D}\mathbf{v}_{Rj}) \qquad (3.8)$$

Writing (3.8) as two matrix equations using all real quantities gives

$$[\overline{\mathbf{K}}\ \overline{\mathbf{D}}]\left[\begin{bmatrix} \mathbf{I} & \mathbf{0} \\ \mathbf{0} & -\omega_j\mathbf{I} \end{bmatrix}\begin{bmatrix} \mathbf{v}_{Rj} \\ \mathbf{v}_{Ij} \end{bmatrix}\begin{bmatrix} \mathbf{0} & \mathbf{I} \\ \omega_j\mathbf{I} & \mathbf{0} \end{bmatrix}\begin{bmatrix} \mathbf{v}_{Rj} \\ \mathbf{v}_{Ij} \end{bmatrix}\right] =$$
$$\left[-\omega_j\mathbf{f}_{Ij} + [-\mathbf{K} + \omega_j^2\mathbf{M}\omega_j\mathbf{D}]\begin{bmatrix} \mathbf{v}_{Rj} \\ \mathbf{v}_{Ij} \end{bmatrix}\omega_j\mathbf{f}_{Rj}+\right.$$
$$\left.[-\omega_j\mathbf{D} - \mathbf{K} + \omega_j^2\mathbf{M}]\begin{bmatrix} \mathbf{v}_{Rj} \\ \mathbf{v}_{Ij} \end{bmatrix}\right] \qquad (3.9)$$

After simplifying notation (3.9) becomes

$$[\overline{\mathbf{K}}\ \overline{\mathbf{D}}][\mathbf{A}_{1j}\hat{\mathbf{v}}_j\mathbf{A}_{2j}\hat{\mathbf{v}}_j] = [-\omega_j\mathbf{f}_{Ij} + \mathbf{B}_{1j}\hat{\mathbf{v}}_j\ \omega_j\mathbf{f}_{Rj} + \mathbf{B}_{2j}\hat{\mathbf{v}}_j] \qquad (3.10)$$

where

$$\hat{\mathbf{v}}_j = \begin{bmatrix} \mathbf{v}_{Rj} \\ \mathbf{v}_{Ij} \end{bmatrix}, \quad \mathbf{A}_{1j} = \begin{bmatrix} \mathbf{I} & \mathbf{0} \\ \mathbf{0} & -\omega_j\mathbf{I} \end{bmatrix}, \quad \mathbf{A}_{2j} = \begin{bmatrix} \mathbf{0} & \mathbf{I} \\ \omega_j\mathbf{I} & \mathbf{0} \end{bmatrix}$$
$$\mathbf{B}_{1J} = [-\mathbf{K} + \omega_j^2\mathbf{M}\omega_j\mathbf{D}], \mathbf{B}_{2j} = [-\omega_j\mathbf{D} - \mathbf{K} + \omega_j^2\mathbf{M}]$$

From (3.10) we get

$$[\overline{\mathbf{K}}\ \overline{\mathbf{D}}]\mathbf{G}_j = \mathbf{V}_j \qquad (3.11)$$

where the matrices in (3.10) and (3.11) correspond, $\mathbf{G}_j$ is $2n \times 2$, and $\mathbf{V}_j$ is $n \times 2$. For $j = 1, 2, \ldots p$ frequency points, equation (3.11) becomes

$$\underset{n\times 2n}{[\overline{\mathbf{K}}\ \overline{\mathbf{D}}]}\underset{2n\times 2p}{[\mathbf{G}_1\mathbf{G}_2\ldots\mathbf{G}_p]} = \underset{n\times 2p}{[\mathbf{V}_1\mathbf{V}_2\ldots\mathbf{V}_p]} \qquad (3.12)$$

or

$$[\overline{\mathbf{K}}\ \overline{\mathbf{D}}]\mathbf{G} = \mathbf{V} \qquad (3.13)$$

where the matrices in (3.12) and (3.13) correspond, and in (3.12) the dimensions are shown below the matrices. The solution to (3.13) must retain the connectivity and symmetry of the FEM matrices and is obtained using an elemental proportional update technique.[43] In this approach, scale factors are defined for each finite element in the model, and these are computed through a linear least squares optimization. The elemental matrices with scale factors are then assembled to obtain the global update matrices. This method exactly preserves the connectivity of the system matrices but assumes that proportional change occurs to the elements, which means, for example, that bending stiffness and extensional stiffness of the redesigned element are dependent.

Alternatively, separate factors could be used to represent the individual stiffnesses. The equations for this approach are as follows. Let

$$\overline{\mathbf{D}} = \sum_{r=1}^{L} d_r^d \mathbf{D}_r, \qquad \overline{\mathbf{K}} = \sum_{r=1}^{L} d_r^k \mathbf{K}_r \qquad (3.14a,b)$$

where the summation denotes matrix assembly of the elemental matrices $\mathbf{K}_r$ and $\mathbf{D}_r$, multiplied by the scale factor coefficients d_r^d and d_r^k, and where L is the total number of finite elements in the model. Rewriting $\mathbf{G}$ as $\mathbf{G} = [\mathbf{G}1^T \ \mathbf{G}2^t]^T$, where $\mathbf{G}1$ and $\mathbf{G}2$ are $n \times 2p$ matrices, and using (3.14), then (3.13) becomes

$$\left(\sum_{r=1}^{L} d_r^k \mathbf{K}_r \right) \mathbf{G}1 + \left(\sum_{r=1}^{L} d_r^d \mathbf{D}_r \right) \mathbf{G}2 = \mathbf{V} \qquad (3.15)$$

Now rewrite (3.15) by stacking the columns of $\mathbf{K}_r \mathbf{G}1$ and $\mathbf{D}_r \mathbf{G}2$ as single columns multiplying the scale factors d_r^d and d_r^k. This procedure produces the expression:

$$\begin{bmatrix} (\mathbf{K}_1\mathbf{G}1)_{col1} & & (\mathbf{K}_L\mathbf{G}1)_{col1} & (\mathbf{D}_1\mathbf{G}2)_{col1} & & (\mathbf{D}_L\mathbf{G}2)_{col1} \\ \vdots & \cdots & & \vdots & \cdots & \\ (\mathbf{K}_1\mathbf{G}1)_{col2p} & & (\mathbf{K}_L\mathbf{G}1)_{col2p} & (\mathbf{D}_1\mathbf{G}2)_{col2p} & & (\mathbf{D}_L\mathbf{G}2)_{col2p} \end{bmatrix}$$
$$\underset{2np \times 2L}{}$$

$$\begin{bmatrix} d_1^k \\ \vdots \\ d_L^d \end{bmatrix} \begin{bmatrix} \mathbf{V}_{col1} \\ \vdots \\ \mathbf{V}_{col2p} \end{bmatrix} \qquad (3.16)$$
$$\underset{2L \times 1}{} \qquad \underset{2np \times 1}{}$$

In some cases to simplify the solution we might assume that damping and stiffness change by the same percent, that is, $d_r^d = d_r^k$. In simpler notation (3.16) becomes

$$\mathbf{Ad} = \mathbf{b} \qquad (3.17)$$

where $\mathbf{d} = [(\mathbf{d}^k)^T \ (\mathbf{d}^d)^T]^T$.

2.3.2. Solution Of The Linear Matrix Equation

Equation (3.17) is generally an overdetermined system of equations and a least-squares solution can be found using various approaches, as follows.

(1) A pseudo-inverse solution can be obtained from $\mathbf{A} = \mathbf{PDQ}$ using the singular value decomposition. The pseudo-inverse of $\mathbf{A}$ is $\mathbf{A}^+ = (\mathbf{Q}^*)^T \mathbf{D}^+ \mathbf{P}^*$, where $\mathbf{P}$ and $\mathbf{Q}$ are unitary matrices and $\mathbf{D}^+$ is diagonal and contains the reciprocals of the singular values or zeros. Then the solution to (3.17) becomes

$$\mathbf{d} = \mathbf{A}^+ \mathbf{b} \qquad (3.18)$$

where $\mathbf{A}^+ = ((\mathbf{A}^*)^T \mathbf{A})^{-1} (\mathbf{A}^*)^T$ if $\mathbf{A}$ is of full rank.

(2) When $\mathbf{A}$ is rank deficient (rank $\mathbf{A} = k \leq 2L$) then the overdetermined problem does not have a unique solution and a possibly better solution for the redesign application is one that will produce the largest number of zero terms in $\mathbf{d}$, rather than the least squares solution which is minimizing norm $|\mathbf{Ad} - \mathbf{b}|$ with minimum norm $|\mathbf{d}|$. Minimizing the number of adjusted terms is realistic as redesign is generally desired over the minimum section of the structure. This second solution with the minimum number of non-zero terms in $\mathbf{d}$ is obtained using the QR factorization[36] as

$$\mathbf{d} = \mathbf{P}(\mathbf{R}\backslash(\mathbf{Q}^T \mathbf{b})) \qquad (3.19)$$

where $\mathbf{P}$, $\mathbf{Q}$ and $\mathbf{R}$ are the permutation, orthogonal and upper triangular matrices obtained from the QR decomposition, and $\backslash$ denotes back substitution using the diagonal $\mathbf{R}$.

(3) If the redesign or control gains are all restricted to be positive, then a non-negative least squares solution will further constrain and often result in a more feasible solution. This solution is given by multiplying (3.17) by -1 to get $\mathbf{A}(-\mathbf{d}) = -\mathbf{b}$, then use the NNLS function in MATLAB[38] as $d = -\text{nnls}(A, -b)$.

(4) In the general case where upper and lower bounds on each of the scale factors is desired, the quadratic programming solution in MATLAB[38, p.3–38] can be used.

(5) If proportional damping is assumed for the structural redesign, then the solution to (3.17) simplifies by replacing $\mathbf{A}$ with $(\mathbf{AK} + \mathbf{AD})$. In this case $\mathbf{d}^d = \mathbf{d}^k$.

An advantage of the algorithm presented is that it requires less compuations that the optimization approach of Section 2. With $p = n$ the solution to (3.13) could be found by direct inversion, and will be unique under the conditions that the frequency response vectors are linearly independent. Practically, this means that the frequency points at which the frequency response vectors are computed must span the range of frequencies where the assigned and original FRFs are significantly different. If the full frequency response vector is measured, ($\mathbf{v}_{1j}$ and $\mathbf{v}_{2j}$) and the FRFs are achievable, then the updated matrices determined from (3.13) will exactly reproduce the assigned FRFs. This formulation is analogous to eigenstructure assignment.

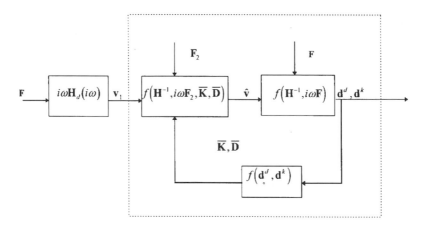

Figure 3. Model For FRF Assignment by Iteration.

However, with incomplete or unassignable vectors the least-squares solution using proportional elemental redesign will be more practical, and the solution to equation (3.17) is generally used.

Since v_{2j} depends on $\overline{\mathbf{D}}$ and $\overline{\mathbf{K}}$ for partial FRF assignment, an improved solution can be obtained by using (3.14) in (3.6) and iterating the procedure. A flow diagram of the iteration technique is shown in Figure 3, where $i\omega\mathbf{H}_d(i\omega)$ represents the mobility FRF assigned to the structure.

2.3.3. Design of an Actively Controlled Bridge Truss

This example assigns a measured frequency response to reduce the vibration amplitude of the 18 member simply supported truss bridge shown in Figure 4. A sinusoidal force $f(t) = 0.01\sin(\omega t)$ acts at nodes 3 and 5, as shown in Figure 4 (DOF's 5,6,9,10 of the analytical model). These forces represent using a Y (90°) pre-tensioned cable with a single force actuator to excite the bridge to model traffic loads. The bridge is constructed of tension-compression truss elements. The mass matrix contains non-structural mass at each bottom node of the bridge. Proportional damping ($\mathbf{D} = 0.001\mathbf{K}$) is assumed for the original structure to simplify the example. Modal damping or any type of damping could also be used. The system mass and stiffness matrices are:

$$\mathbf{M} = 0.0005(\rho A/l)$$
$$\text{diag}([8.414; 8.414; 3.0; 3.0; 8.414; 8.414; 5.828; 5.828; 8.414; 8.414;}$$
$$5.828; 5.828; 8.414; 8.414; 3.0; 3.0)$$

$$\mathbf{K} = (EA/l)$$

$$
\begin{bmatrix}
2354 & 0.354 & 0 & 0 & -1 & 0 & -0.354 & -0.354 & 0 & 0 & 0 & 0 & 0 & 0 & 0 & 0 \\
 & 1.354 & 0 & -1 & 0 & 0 & -0.354 & -0.354 & 0 & 0 & 0 & 0 & 0 & 0 & 0 & 0 \\
 & & 1.354 & 0.354 & 0 & 0 & -1 & 0 & 0 & 0 & 0 & 0 & 0 & 0 & 0 & 0 \\
 & & & 1.354 & 0 & 0 & 0 & 0 & 0 & 0 & 0 & 0 & 0 & 0 & 0 & 0 \\
 & & & & 2.354 & 0.354 & 0 & 0 & -1 & 0 & -0.354 & -0.354 & 0 & 0 & 0 & 0 \\
 & & & & & 1.354 & 0 & -1 & 0 & 0 & -0.354 & -0.354 & 0 & 0 & 0 & 0 \\
 & & & & & & 2.708 & 0 & -0.354 & 0.354 & -1 & 0 & 0 & 0 & 0 & 0 \\
 & & & & & & & 1.708 & 0.354 & -0.354 & 0 & 0 & 0 & 0 & 0 & \\
 & & & & & & & & 2.354 & -0.354 & 0 & 0 & -1 & 0 & 0 & 0 \\
 & & & & & & & & & 1.354 & 0 & -1 & 0 & 0 & 0 & 0 \\
 & & & & & & & & & & 2.708 & 0 & -0.354 & 0.354 & -1 & 0 \\
 & \text{symm} & & & & & & & & & & 1.708 & 0.354 & -0.354 & 0 & 0 \\
 & & & & & & & & & & & & 2.354 & -0.354 & 0 & 0 \\
 & & & & & & & & & & & & & 1.354 & 0 & -1 \\
 & & & & & & & & & & & & & & 1.354 & -0.354 \\
 & & & & & & & & & & & & & & & 1.354
\end{bmatrix}
$$

where ρ, A, l, and E are the mass density, area, length and modulus of elasticity of the truss members. The vector of displacements is $\mathbf{x}(t) = [x1y1x2y2\ldots x16y16]^T$. The structural connectivity information is given in[35]. The first undamped natural frequency in the original model is 0.516 Hz. The redesign algorithm developed uses the elemental connectivity information and automatically reorders the system equations into the 1 and 2 partitions. This allows the structure to be efficiently reanalyzed, and also a large number of redesign cases can be tested quickly. The node and element numbers are given in Figure 4.

A difficult aspect of the FRF assignment procedure is defining the FRFs to assign to the model. Just as eigenstructure must lie in an achievable subspace to be assignable, only a certain class of FRF vectors can be assigned. To assure the assignability of the FRF vectors, in this example simulated data is obtained by putting a known proportional damping in the analytical model and computing the mobility frequency response to assign to the original model. The proportional damping factor used for the simulation is $(\mathbf{D} = 0.02\mathbf{K})$, which is an increase of 20 times over the original damping. Since a proportional damping matrix with large damping, in practice, cannot be achieved, the FRFA algorithm will be used to try to assign the highly damped FRFs using a minimum number of changes to the structure. The simulated FRF is assigned at all nodes in the vertical direction (Y axis), that is DOFs, 2, 4, 6, 8, 10, 12, 14, 16, in the frequency range from 0.5 to 50 rad/sec. When the FRF assignment to the original model is exact, the FRF of the updated model moves on top of the FRF of the assigned model.

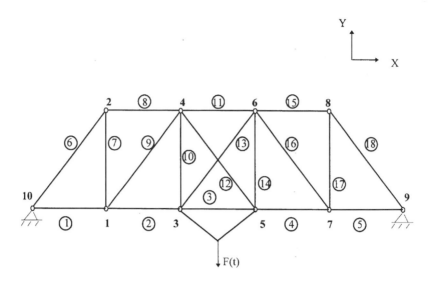

Figure 4. Finite Element Model of a Truss Bridge with Excitation Force.

The general design procedure for the active structure is given as follows in three steps.

(S1) Apply a large proportional or other damping to the original lightly damped uncontrolled structure. Calculate the FRFs that are to be assigned to the model.

(S2) Assign the FRFs from S1 to the original model. The dk and dd vectors define the changes to the model required to assign the FRFs.

(S3) Ignore all small changes in dk and dd defined in S2. Re-run the assignment using only the most significant changes. If the new dk and dd vectors are acceptable and the assignment is close to desired, then the new design is considered satisfactory. Otherwise, more extensive changes (S2) or new FRFs (S1) should be tried.

In the bridge truss problem there are 18 members of the truss. General damping was assumed, thus there are 36 design variables in the optimization. Performing step 1 resulted in highly damped FRFs to assign to the structure. Assigning these in step 2 in the vertical direction produced updated FRF's in 12 iterations that were close but did not exactly match the assigned FRFs. This is because we are trying to match the assigned FRFs in the vertical direction only using a minimum number of changes to the structure, that is, a least squares solution. The resulting coefficients in order for the 18 elements in the model are: **dk** = [.0 .0017 .0 .0017 .0 .0 .3088 .0007 .0 .1866 .0 .0 .0 .1866 .0007 .0 .3088 .0], **dd** = [.0 .0 .0 .0 .0 1.2645 41.3987 .4608 4.1848 18.6272 .3681 .0 .0 18.6272 .4608 4.1848 41.3987 1.2645]. Note that the changes are symmetric, and the number of design variables could have been reduced by one half based on the symmetry of the model and loading. Since a number of values are small they are neglected and the assignment is re-run. The coefficients chosen for step three are: **dk** = [.0, .0, .0, .0, .0, .0, .3088, .0, .0, .1866, .0, .0, .0, .1866, .0, .0, .3088, .0], **dd** = [.0, .0, .0, .0, .0, .0, 41.4, .0, .0, 18.6, .0, .0, .0, 18.6, .0, .0, 41.4, .0]. Further iteration did not significantly change these coefficients. Thus these coefficients are added to the original values of the design variables that are all 1 for the original model and the updated design vectors are: **Kd** = [1, 1, 1, 1, 1, 1, 1.3088, 1, 1, 1.1866, 1, 1, 1, 1.1866, 1, 1, 1.3088, 1], **Dd** = [1, 1, 1, 1, 1, 1, 42.4, 1, 1, 19.6, 1, 1, 1, 19.6, 1, 1, 42.4, 1]. This result specifies that we must increase the stiffness of elements 7 and 17 by 30.88%, and elements 10 and 14 by 18.66%. We must also increase the damping of elements 7 and 17 by a factor of 42.4, and elements 10 and 14 by a factor of 19.6. The FRFs for DOFs 2, 6, 8 (*Y* axis), and 1, 5, 7 (*x* axis) are plotted in Figure 5. The peak values of these plots are given in Table 1. In the Y direction where the amplitudes are large, the redesign reduced the peaks of the mobility FRFs by 89%. The eigenvalues of the original and the redesigned models are given below.

$$
\lambda_{\text{original}} =
\begin{bmatrix}
-.0052 - 3.23951i \\
-.0241 - 6.9441i \\
-.0474 - 9.7372i \\
-.0539 - 10.3780i \\
-.0796 - 12.6181i \\
-.1799 - 18.9665i \\
-.1893 - 19.4570i \\
-.2509 - 22.4002i \\
-.3011 - 24.4002i \\
-.4020 - 28.3526i \\
-.4258 - 29.1789i \\
-.4914 - 31.3454i \\
-.5374 - 32.7808i \\
-.5899 - 34.3443i \\
-.6798 - 36.8660i \\
-.8258 - 40.6307i
\end{bmatrix}
\quad
\lambda_{\text{redesigned}} =
\begin{bmatrix}
-.1235 - 3.7771i \\
-.1805 - 7.2828i \\
-.2447 - 10.0281i \\
-2.1049 - 12.9790i \\
-1.6757 - 14.9997i \\
-.2379 - 19.3803i \\
-1.1887 - 21.4010i \\
-9.1412 - 22.4045i \\
-1.4159 - 26.3548i \\
-18.0304 - 25.0780i \\
-1.0906 - 31.0242i \\
-2.0051 - 31.4198i \\
-1.0163 - 31.5677i \\
-6.5809 - 31.5515i \\
-8.6684 - 37.0834i \\
-1.8388 - 38.1995i
\end{bmatrix}
$$

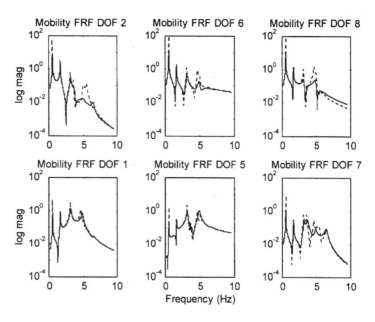

Figure 5. Frequency Response Functions for uncontrolled structure (dashed line) and redesigned active structure (solid line).

Referring to (3.1), the update stiffness matrix is given as:

$$
\overline{\mathbf{K}} =
\begin{bmatrix}
0 & 0 & 0 & 0 & 0 & 0 & 0 & 0 & 0 & 0 & 0 & 0 & 0 & 0 & 0 & 0 \\
 & .3088 & 0 & -.3088 & 0 & 0 & 0 & 0 & 0 & 0 & 0 & 0 & 0 & 0 & 0 & 0 \\
 & & 0 & 0 & 0 & 0 & 0 & 0 & 0 & 0 & 0 & 0 & 0 & 0 & 0 & 0 \\
 & & & .3088 & 0 & 0 & 0 & 0 & 0 & 0 & 0 & 0 & 0 & 0 & 0 & 0 \\
 & & & & 0 & 0 & 0 & 0 & 0 & 0 & 0 & 0 & 0 & 0 & 0 & 0 \\
 & & & & & .1866 & 0 & -.1866 & 0 & 0 & 0 & 0 & 0 & 0 & 0 & 0 \\
 & & & & & & 0 & 0 & 0 & 0 & 0 & 0 & 0 & 0 & 0 & 0 \\
 & & & & & & & .1866 & 0 & 0 & 0 & 0 & 0 & 0 & 0 & 0 \\
 & & & & & & & & 0 & 0 & 0 & 0 & 0 & 0 & 0 & 0 \\
 & & & & & & & & & .1866 & 0 & -.1866 & 0 & 0 & 0 & 0 \\
 & \textit{Symm} & & & & & & & & & 0 & 0 & 0 & 0 & 0 & 0 \\
 & & & & & & & & & & & .1866 & 0 & 0 & 0 & 0 \\
 & & & & & & & & & & & & 0 & 0 & 0 & 0 \\
 & & & & & & & & & & & & & .3088 & 0 & -.3088 \\
 & & & & & & & & & & & & & & 0 & 0 \\
 & & & & & & & & & & & & & & & .3088
\end{bmatrix}
$$

Table 1. Comparison of peak values of FRFs for original and redesigned models

Node and Direction	Magnitude of FRF Original Model	Magnitude of FRF Redesigned Model	% Reduction in Peak of FRF
1–X	4.4521	1.3270	70.2
1–Y	57.6720	6.1811	89.3
3–X	2.4274	1.0957	54.9
3–Y	95.2392	10.3575	89.1
4–X	8.5589	0.9526	88.9
4–Y	90.3572	9.8667	89.1

The update damping matrix is given as:

$$
\overline{\mathbf{D}} =
\begin{bmatrix}
0 & 0 & 0 & 0 & 0 & 0 & 0 & 0 & 0 & 0 & 0 & 0 & 0 & 0 & 0 & 0 \\
 & .0.414 & 0 & -.0414 & 0 & 0 & 0 & 0 & 0 & 0 & 0 & 0 & 0 & 0 & 0 & 0 \\
 & & 0 & 0 & 0 & 0 & 0 & 0 & 0 & 0 & 0 & 0 & 0 & 0 & 0 & 0 \\
 & & & .0414 & 0 & 0 & 0 & 0 & 0 & 0 & 0 & 0 & 0 & 0 & 0 & 0 \\
 & & & & 0 & 0 & 0 & 0 & 0 & 0 & 0 & 0 & 0 & 0 & 0 & 0 \\
 & & & & & .0186 & 0 & -.0186 & 0 & 0 & 0 & 0 & 0 & 0 & 0 & 0 \\
 & & & & & & 0 & 0 & 0 & 0 & 0 & 0 & 0 & 0 & 0 & 0 \\
 & & & & & & & .0186 & 0 & 0 & 0 & 0 & 0 & 0 & 0 & 0 \\
 & & & & & & & & 0 & 0 & 0 & 0 & 0 & 0 & 0 & 0 \\
 & & & & & & & & & .0186 & 0 & -.0186 & 0 & 0 & 0 & 0 \\
 & & & \text{Symm} & & & & & & & 0 & 0 & 0 & 0 & 0 & 0 \\
 & & & & & & & & & & & .0186 & 0 & 0 & 0 & 0 \\
 & & & & & & & & & & & & 0 & 0 & 0 & 0 \\
 & & & & & & & & & & & & & .0414 & 0 & -.0414 \\
 & & & & & & & & & & & & & & 0 & 0 \\
 & & & & & & & & & & & & & & & .0414
\end{bmatrix}
$$

The stiffness changes specified by the redesign may be made by changing the thickness or material of the truss members. However, the damping changes would require discrete viscous dampers parallel to elements 7, 17 and 10, 14. Alternatively, active damping control could be used to obtain the desired damping force. The control law for the collocated active damping design is given as follows. Write the closed-loop equation of motion as: $\mathbf{M\ddot{x}} + \mathbf{D\dot{x}} + \mathbf{Kx} = \mathbf{f} + \mathbf{Bu}$ where $\mathbf{u}_v = -\mathbf{F}_v\dot{\mathbf{y}}$, and $\dot{\mathbf{y}} = \mathbf{C\dot{x}}$. The control influence matrix for four actuators at elements 7, 10, 14, 17 is:

$$
\mathbf{B} =
\begin{bmatrix}
0 & 1 & 0 & -1 & 0 & 0 & 0 & 0 & 0 & 0 & 0 & 0 & 0 & 0 & 0 & 0 \\
0 & 0 & 0 & 0 & 0 & 1 & 0 & -1 & 0 & 0 & 0 & 0 & 0 & 0 & 0 & 0 \\
0 & 1 & 0 & -1 & 0 & 0 & 0 & 0 & 0 & 0 & 0 & -1 & 0 & 0 & 0 & 0 \\
0 & 1 & 0 & -1 & 0 & 0 & 0 & 0 & 0 & 0 & 0 & 0 & 0 & 1 & 0 & -1
\end{bmatrix}^T
$$

Since we are using collocated control, the measurement location matrix is
$\mathbf{C} = \mathbf{B}^T$. The update active damping matrix is $\overline{\mathbf{D}} = \mathbf{B}\mathbf{F}_v\mathbf{B}^T$. The gain matrix
is obtained from $\mathbf{F}_v = (\mathbf{B}^T\mathbf{B})^{-1}(\mathbf{B}^T\overline{\mathbf{D}}\mathbf{B})(\mathbf{B}^T\mathbf{B})$ and is shown below for this
example as:

$$
\mathbf{F}_v = \begin{bmatrix} .0414 & 0 & 0 & 0 \\ 0 & .0186 & 0 & 0 \\ 0 & 0 & .0186 & 0 \\ 0 & 0 & 0 & .0414 \end{bmatrix}
$$

The redesign by increasing stiffness in certain elements and adding ac-
tive/passive discrete dampers theoretically reduces structural vibration at the
first resonance point by 89% without significantly changing the lower natural
frequencies of the structure.

2.3.4. Section Summary

The FRFA technique uses FRF information and can design an active structure
to meet vibration requirements within a desired frequency range. A least-
squares solution with iteration is used which is more efficient than a nonlinear
optimization approach.

2.4. DESIGN OF NONLINEAR STRUCTURES AND CONTROLLERS

Advanced structures such as the space station[44] and future high speed
aircraft[33] are being designed for minimum weight to improve performance
and reduce operating costs over the life of the product. These light-weight
designs are more flexible than traditional structures and behave nonlinearily,
and have a finite fatigue life that is determined using Miner's rule and
the stress to number of cycles to failure curves (S-N curves) for the
structural materials. For most engineering materials, these logarithmic curves
show that a small number of fatigue cycles at large stress levels can
cause greater damage than a large number of cycles at lower stresses.
Performance requirements of these flexible structures also dictate that
vibration is suppressed quickly. Thus it is very important in the design of
flexible structures to suppress vibrations and prevent large displacements
and stresses. To design efficient structures and to accurately predict structural
life requires that highly flexible structures be modeled and analyzed using
nonlinear structural theory, and since increased flexibility increases vibration,
these structures must also be designed with some type of passive or active
vibration suppression devices. Therefore an appropriate design tool for
flexible structures must use a nonlinear model and should also simultaneously

design a vibration controller. This approach will accurately predict and limit structural responses and will be faster than performing separate structure and controller design procedures, and hence will improve the safety, reliability, and performance of light-weight structures.

Although nonlinear design is advocated to develop light-weight low cost structures, presently there are few general techniques available to simultaneously design the structure and controller for nonlinear systems. Analytical techniques to optimize nonlinear structures and controllers are difficult to develop because exact closed-form solutions to general large order systems of nonlinear differential equations are almost impossible to obtain, and approximation techniques can lead to suboptimal designs.

Existing techniques for the control of nonlinear structures include feedback linearization, describing functions, sliding control, and adaptive control[45], model referenced adaptive control[46], neural network control[47], nonlinear modal control[48], and H_∞ control.[49-51] These approaches all have limitations when applied to the control of large flexible structures including restriction to certain classes of nonlinear systems or only linear systems, requiring full state feedback or an observer, errors due to truncating sub-harmonic and super-harmonic Fourier terms, linearization errors, application only to tracking control or uncertain systems, restriction to very small order or single input or decoupled systems, or being heuristically based. These control techniques are also difficult to couple to a multi-disciplinary optimization of the structure. Direct Optimal Control of Nonlinear Systems via Hamiltons Law of Varying Action[52] using assumed time modes can design controllers for nonlinear structures, but the optimal control technique does not use output feedback and is not developed for constrained structural optimization, and requires a lot of computation to obtain the steady-state response for large systems.

This section presents an automated design technique for large nonlinear or linear flexible structures[53,54] and controllers. The advantage of this technique is that explicit time integration is used to take advantage of the sparsity of the nonlinear dynamic and control equations to reduce computations and to allow linear algebraic gradient equations to be developed that can be efficiently solved. Since full system sparse equations in second order form are used in the design, linearization errors, eigenvalue/vector calculations, reduced basis approximations, and spillover effects present in linear design techniques are all eliminated.

2.4.1. Nonlinear Design Technique

The design approach is to build a nonlinear structural model using nonlinear finite elements and use explicit time integration to solve the simultaneous second-order nonlinear differential equations describing the closed loop system. The structure and control optimization is performed by defining an objective function to drive the forced nonlinear system to its zero equilibrium solution using the minimum control force. Constant gain output feedback control is used, which is the simplest and most practical technique for active structural control. No observer or dynamic compensator is needed and time delay in computing the control action is minimized. The control gains are actually the coefficients of some polynomial function of the nonlinear displacements and velocities of the system. The technique can also be used to study the dynamic behavior and control order characteristics of nonlinear systems. Gain suppression can be used to identify the best out of a group of candidate locations to take feedback measurements and to place actuators. Any type of forcing functions can be used and chaotic responses can also be computed.

Nonlinear structural optimization is computationally intensive due to the recomputation of the response at each iteration of the optimization. However, since the sparsity of the structural matrices is preserved in this approach, the new sparse matrix solvers recently available can be used to greatly reduce this computational burden. New parallel-vector algorithms also exist to perform general matrix computations[55], assemble finite element matrices from the elemental level[56], compute design sensitivities[57], and perform large scale design optimization.[58,59] Therefore the method developed here can be parallelized to handle large engineering problems.

Nonlinear systems can have bifurcations to a higher response level, period doubling leading to chaotic behavior, extreme sensitivity to small variations in system and excitation parameters and initial conditions, and internal resonances. The technique presented here can also be used to accurately study whether active control can suppress or take advantage these effects for large structures.

2.4.2. Integration Solution

The general form of the differential equations of motion that describe an n dimensional second order closed-loop nonlinear dynamic system subject to a general forcing function is

$$\mathbf{M}\ddot{\mathbf{x}} + \mathbf{f}_v(\dot{\mathbf{x}}) + \mathbf{f}_p(\mathbf{x}) = \mathbf{F}(t) - \mathbf{u}_v(\dot{\mathbf{x}}) - \mathbf{u}_p(\mathbf{x}) \qquad (4.1)$$

where $\mathbf{x}$ is the displacement vector, $\mathbf{M}$ is the mass matrix, $\mathbf{f}_v(\dot{\mathbf{x}})$ and $\mathbf{f}_p(\mathbf{x})$ are the nonlinear damping and stiffness force vectors, respectively, $\mathbf{F}(t)$ is the external force vector, and $-\mathbf{u}_v(\dot{\mathbf{x}}) - \mathbf{u}_p(\mathbf{x})$ are the nonlinear control force vectors. The external force can be of any form, transient or steady-state. The

Newmark-Beta explicit time integration technique is used to solve (4.1). The solution is:

$$\mathbf{x}_{r+1} = \mathbf{x}_r + \dot{\mathbf{x}}_r \Delta t + [(0.5 - \beta)\ddot{\mathbf{x}}_r + \beta \ddot{\mathbf{x}}_{r+1}]\Delta t^2 \qquad (4.2a)$$

$$\dot{\mathbf{x}}_{r+1} = \dot{\mathbf{x}}_r + [(1 - \lambda)\ddot{\mathbf{x}}_r + \lambda \ddot{\mathbf{x}}_{r+1}]\Delta t \qquad (4.2b)$$

$$\ddot{\mathbf{x}}_{r+1} = \mathbf{M}^{-1}[\mathbf{F}((r+1)\Delta t) - \mathbf{u}_v(\dot{\mathbf{x}}_{r+1}) - \mathbf{u}_p(\mathbf{X}_{r+1}) - \mathbf{f}_v(\dot{\mathbf{x}}_{r+1}) - \mathbf{f}_p(\mathbf{x}_{r+1})]$$
$$(4.2c)$$

where $r + 1$ denotes the current solution point, r denotes the last solution point, Δt is the time step, and λ and β are constants. The integration solution is carried out only long enough to represent the essential characteristics of the response, which may be only four cycles of response. An iteration of (4.2a-c) is done each timestep to obtain an equilibrium force balance that greatly improves the accuracy of the solution. The equilibrium balance is done by taking the first estimate of $\ddot{\mathbf{x}}_{r+1} = \ddot{\mathbf{x}}_r$ and using (4.2a, 4.2b) to get an updated estimate of $\ddot{\mathbf{x}}_{r+1}$ in (4.2c). Note that all terms with the r subscript do not change in this iteration. The value of $\mathbf{x}_{r+1}$ usually converges within about six iterations. The only inversion in the solution is the mass matrix, and the sparsity of the structural matrices is always preserved. Since $\mathbf{M}$ is often diagonal or constant, the inversion is either trivial or only needs to be done once. Furthermore, note that the stiffness matrix is nonlinear but here no repeated inversion is needed to integrate the nonlinear equations. This is in contrast to implicit integration techniques that use repeated inversion of the stiffness matrix (recursion) to obtain the response. The explicit Newmark-Beta integration method is used because it is more accurate than the Euler finite difference method, and it is possible to derive analytic gradients and preserve the sparsity of the structural matrices, which would be difficult to do using the Runge-Kutta or implicit Newmark-Beta methods. A small time-step is needed for the explicit integration, but it is often more efficient to perform many simple calculations (matrix multiplications) rather than fewer complex calculations (matrix inversions). A larger time step can be used in the integration for preliminary design to reduce computations, but accuracy is reduced and stability cannot be guaranteed for nonlinear systems.

2.4.3. Objective Function and Gradient

An objective function that minimizes the mean square vibration displacement of the system and the mean square control forces is

$$J = \frac{1}{np} \sum_{r=1}^{np} (\mathbf{x}_r^T \mathbf{Q}_p \mathbf{x}_r + \dot{\mathbf{x}}_r^T \mathbf{Q}_v \dot{\mathbf{x}}_r + \mathbf{u}_v^T \mathbf{R}_v \mathbf{u}_v + \mathbf{u}_p^T \mathbf{R}_p \mathbf{u}_p) \qquad (4.3)$$

where T denotes transpose, $\mathbf{Q}_p$, $\mathbf{Q}_v$, $\mathbf{R}_v$, and $\mathbf{R}_p$ are diagonal positive semi-definite weighting matrices, and np is the number of solution points used to define the response of the system. For damping design cases, it is necessary to include $\dot{\mathbf{x}}_r^T \mathbf{Q}_v \dot{\mathbf{x}}_r$ in Equation (4.3) to ensure the stability of optimization process. The design variables in the problem are any desired structural parameters and control gains. A standard optimization routine, such as contained within the MATLAB optimization toolbox[38], can be used to minimize Equation (4.3). The only constraints on the design variables will be bounds on their magnitudes that are given as follows

$$\xi_j^L \le \xi_j \le \xi_j^U \qquad j = 1, 2, \ldots, ndv \tag{4.4}$$

where ξ_j are the design variables, ξ_j^L and ξ_j^U represent the lower and upper bounds, and ndv is the total number of design variables in the problem. No functional constraints are needed with this formulation. A final and very important requirement to make the optimization computationally feasible is to derive a closed form gradient of the objective function. As shown below, the gradient of the control gains and structural variables for simple elements can be obtained exactly without any additional function evaluations or recursion that is normally required for nonlinear systems.

The closed-form gradient of the objective function is

$$\frac{dJ}{d\xi_j} = \frac{1}{np} \sum_{r=1}^{np} \left[2\mathbf{x}_r^T \mathbf{Q}_p \frac{\partial \dot{\mathbf{x}}}{\partial \xi_j} + 2\dot{\mathbf{x}}_r^T \mathbf{Q}_v \frac{\partial \dot{\mathbf{x}}}{\partial \xi_j} + 2\mathbf{u}_v^T \mathbf{R}_v \frac{\partial \mathbf{u}_v}{\partial \xi_j} + 2\mathbf{u}_p^T \mathbf{R} \frac{\partial \mathbf{u}_p}{\partial \xi_j} \right] \tag{4.5}$$

Computation of the sensitivity vectors in Equation (4.5) depends upon the form of the equations and can be done closed-form or semi-analytically, depending on the type of finite element used and the type of parametization for the shape optimization of the structure, if used.

The sensitivity vectors are computed by considering the gradient of the displacements. Equation (4.2a) defines the value of x for the current time step $r + 1$. The gradient $\mathbf{X}_{r+1}$ of with respect to the structural-control design variables is obtained by first computing the gradients of Equations (4.2), which are:

$$\frac{\partial \mathbf{x}_{r+1}}{\partial \xi_j} = \left[\frac{\partial \mathbf{x}_r}{\partial \xi_j} + \frac{\partial \dot{\mathbf{x}}_r}{\partial \xi_j} \Delta t + (0.5 - \beta)(\Delta t)^2 \frac{\partial \ddot{\mathbf{x}}_r}{\partial \xi_j} \right] + \beta (\Delta t)^2 \frac{\partial \ddot{\mathbf{x}}_{r+1}}{\partial \xi_j} \tag{4.6a}$$

$$\frac{\partial \dot{\mathbf{x}}_{r+1}}{\partial \xi_j} = \left[\frac{\partial \dot{\mathbf{x}}_r}{\partial \xi_j} + (1 - \lambda)\Delta t \frac{\partial \ddot{\mathbf{x}}_r}{\partial \xi_j} \right] + \lambda \Delta t \frac{\partial \ddot{\mathbf{x}}_{r+1}}{\partial \xi_j} \tag{4.6b}$$

$$\frac{\partial \ddot{\mathbf{x}}_{r+1}}{\partial \xi_j} = -\mathbf{M}^{-1} \left[\frac{\partial \mathbf{u}_v(\dot{\mathbf{x}}_{r+1})}{\partial \xi_j} + \frac{\partial \mathbf{u}_p(\mathbf{x}_{r+1})}{\partial \xi_j} + \frac{\partial \mathbf{f}_v(\dot{\mathbf{x}}_{r+1})}{\partial \xi_j} \right.$$
$$\left. + \frac{\partial \mathbf{f}_p(\mathbf{x}_{r+1})}{\partial \xi_j} \right] \tag{4.6c}$$

Note that all terms in Equations (4.6a) and (4.6b) with the r subscript are known from the previous time step. Thus, substituting Equations (4.6a) and (4.6b) into (4.6c) initially gives a linear implicit Equation that is then rearranged and an inversion is performed to get $\partial \ddot{\mathbf{x}}_{r+1}/\partial \xi_j$ exactly. The gradient of the acceleration Equation (4.6c) is then substituted into Equations (4.6a, 4.6b) to get $\partial \dot{\mathbf{x}}_{r+1}/\partial \xi_j$ and $\partial \mathbf{x}_{r+1}/\partial \xi_j$, and these are substituted into (4.5) to get the gradient of the objective function. The gradient of the control forces is derived using the same procedure to derive Equations (4.6). The solution of the dynamic equations and gradients are calculated in a running fashion so that only the current and last time step solution information is saved. This reduces computations and memory requirements. As stated, the gradient equations are initially linear equations in an implicit form, but they are rearranged and solved by inversion rather than recursion that is normally needed when using a harmonic balance type solution for nonlinear differential equations. This is possible because the displacements are known and can be directly substituted into the nonlinear terms thus eliminating recursion. The formulation presented above represents a general approach to analyze nonlinear actively controlled structures, and lessens the major difficulties inherent with nonlinear design, that is, intractable computations, and non-optimal solutions to the optimization problem.

2.4.4. Design and Control of a Nonlinear Truss

The traditional way to describe a flexible structure is to use the finite element method (FEM), which represents each element of the structure separately with continuity at common nodes, and the total structure is approximated by linear equations at the nodes. The FEM linear equations can be used to analyze static or dynamic behavior, and this is sufficiently accurate for most applications. Since the FEM model is linear, active control can be designed according to linear system theory such as classical control methods and modern control theory that are based upon transfer functions or linear state space equations. Unfortunately, practical structures may be nonlinear due to geometric or material nonlinearity, or vibrations outside the linear range due to large external forces, and then the linear design techniques are not accurate. In general, for large flexible structures, a nonlinear model can more accurately describe the motion of the structure, and therefore a more accurate controller can be designed based on the nonlinear structural model. Thus it is useful to develop nonlinear system theories and techniques to design flexible structures.

In this section, the nonlinear design technique is used to optimize the design of a four bay 16 degree-of-freedom cantilever truss with a transverse

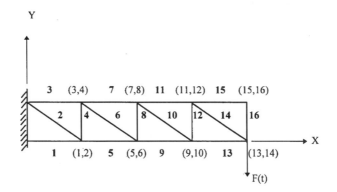

Figure 6. Nonlinear truss finite element model (note: (i, j) represents x, y DOF at each node).

harmonic force at its end, and controlled using a single actuator near the support. The truss structure is shown in Figure 6, and the undeformed and deformed configuration of a truss member are shown in Figure 7. All members of the truss are pin connected. The design goal is to reduce the cantilever type bending of the truss.

A new geometrically-exact truss finite element has been derived using Jaumann strains.[60–61] The truss shown in Figure 6 is modeled using this new finite element formulation, which is given below. The dynamic version of the principle of virtual work states that

$$\int (\delta K_e - \delta \pi + \delta W_{nc}) dt = 0 \tag{4.7}$$

where π = elastic emergy; K_e = kinetic energy; and W_{nc} = nonconservative energy including the energy due to external distributed and/or concentrated loads and dampings. Here it is assumed that K_e and W_{nc} have the same characteristics as a linear model of the truss. In the following we will discuss the elastic energy π which accounts for the geometric nonlinearities. Referring to Figure 7, the elemental elastic energy can be derived as

$$\delta \pi^{(e)} = \int \sigma_{11} \delta \varepsilon_{11} dx A \tag{4.8}$$

where A is the cross-sectional area of the member. The relationship between Jaumann stress and Jaumann strain is $\sigma_{11} = E \varepsilon_{11}$ where E is the Youngs Modulus of the elemental material. Assuming that ε_{11} is a constant within

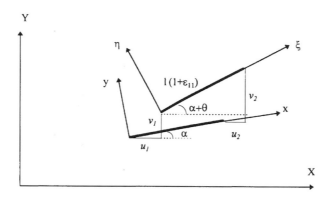

Figure 7. Nonlinear finite element description.

a two-force member and letting $\phi_i = \frac{\partial \varepsilon_{11}}{\partial u_i}$, then Equation (4.8) can be expanded into

$$\delta \pi^{(e)} = \int E\varepsilon_{11}\delta\varepsilon_{11}dxA$$

$$= EAl\varepsilon_{11}\delta\varepsilon_{11}$$

$$= EAl\varepsilon_{11}(\phi 1\delta u1 + \phi 2\delta u2 + \phi 3\delta u3 + \phi 4\delta u4)$$

$$= \{\delta\mathbf{U}^{(e)}\}^T \{\mathbf{KU}^{(e)}\} \qquad (4.9)$$

where $\{\mathbf{U}^{(e)}\}$ is the elemental nodal displacement vector given as

$$\{\mathbf{U}^{(e)}\} = \{u_1, u_2, u_3, u_4\}^T = \{u_1, v_1, u_2, v_2\}^T$$

Therefore, similar to the linear finite element method, we have the local stiffness term

$$\{\mathbf{Ku}^{(e)}\} = \begin{bmatrix} g_1^{(e)} \\ g_2^{(e)} \\ g_3^{(e)} \\ g_4^{(e)} \end{bmatrix} = \begin{bmatrix} EAl\varepsilon 11\phi 1 \\ EAl\varepsilon 11\phi 2 \\ EAl\varepsilon 11\phi 3 \\ EAl\varepsilon 11\phi 4 \end{bmatrix} \qquad (4.10)$$

where

$$\varepsilon_{11} = \beta - 1 \qquad (4.11)$$

$$\beta = \sqrt{\left(\frac{u2 - u1}{l} + \cos\alpha\right)^2 + \left(\frac{v2 - v1}{l} + \sin\alpha\right)^2}$$

$$\phi1 = \frac{\partial\varepsilon_{11}}{\partial u_1} = \frac{-1}{\beta l}\left(\frac{u2 - u1}{l} + \cos\alpha\right)$$

$$\phi2 = \frac{\partial\varepsilon_{11}}{\partial v_1} = \frac{-1}{\beta l}\left(\frac{v2 - v1}{l} + \sin\alpha\right)$$

$$\phi3 = \frac{\partial\varepsilon_{11}}{\partial u_1} = \frac{1}{\beta l}\left(\frac{u2 - u1}{l} + \cos\alpha\right)$$

$$\phi4 = \frac{\partial\varepsilon_{11}}{\partial v_2} = \frac{1}{\beta l}\left(\frac{v2 - v1}{l} + \sin\alpha\right)$$

The second derivatives of strain (11 (i.e., the first derivatives of (i) with respect to displacements at every DOF are used for the gradient calculation in the application of the nonlinear design technique. They are:

$$\frac{\partial^2}{\partial u_1 \partial u_1}\varepsilon_{11} = \frac{1}{\beta_l^2} + \left(\frac{u2 - u1}{l} + \cos\alpha\right)\frac{1}{l\beta^3}\left(\frac{u2 - u1}{l} + \cos\alpha\right)\frac{-1}{l}$$

$$\frac{\partial^2}{\partial u_1 \partial v_1}\varepsilon_{11} = \left(\frac{u2 - u1}{l} + \cos\alpha\right)\frac{1}{l\beta^3}\left(\frac{v2 - v1}{l} + \sin\alpha\right)\frac{-1}{l}$$

$$\frac{\partial^2}{\partial u_1 \partial u_2}\varepsilon_{11} = \frac{-1}{\beta_l^2} + \left(\frac{u2 - u1}{l} + \cos\alpha\right)\frac{1}{l\beta^3}\left(\frac{u2 - u1}{l} + \cos\alpha\right)\frac{1}{l}$$

$$\frac{\partial^2}{\partial u_1 \partial v_2}\varepsilon_{11} = \left(\frac{u2 - u1}{l} + \cos\alpha\right)\frac{1}{l\beta^3}\left(\frac{v2 - v1}{l} + \sin\alpha\right)\frac{1}{l}$$

$$\frac{\partial^2}{\partial v_1 \partial v_1}\varepsilon_{11} = \frac{1}{\beta_l^2} + \left(\frac{v2 - v1}{l} + \sin\alpha\right)\frac{1}{l\beta^3}\left(\frac{v2 - v1}{l} + \sin\alpha\right)\frac{-1}{l}$$

$$\frac{\partial^2}{\partial v_1 \partial u_2}\varepsilon_{11} = \left(\frac{v2 - v1}{l} + \sin\alpha\right)\frac{1}{l\beta^3}\left(\frac{u2 - u1}{l} + \cos\alpha\right)\frac{1}{l}$$

$$\frac{\partial^2}{\partial v_1 \partial v_2}\varepsilon_{11} = \frac{-1}{\beta_l^2} + \left(\frac{v2 - v1}{l} + \sin\alpha\right)\frac{1}{l\beta^3}\left(\frac{v2 - v1}{l} + \sin\alpha\right)\frac{1}{l}$$

$$\frac{\partial^2}{\partial u_2 \partial u_2}\varepsilon_{11} = \frac{1}{\beta_l^2} + \left(\frac{u2 - u1}{l} + \cos\alpha\right)\frac{1}{l\beta^3}\left(\frac{u2 - u1}{l} + \cos\alpha\right)\frac{-1}{l}$$

$$\frac{\partial^2}{\partial u_2 \partial v_2}\varepsilon_{11} = \left(\frac{u2 - u1}{l} + \cos\alpha\right)\frac{1}{l\beta^3}\left(\frac{v2 - v1}{l} + \sin\alpha\right)\frac{-1}{l}$$

$$\frac{\partial^2}{\partial v_2 \partial v_2}\varepsilon_{11} = \frac{1}{\beta_l^2} + \left(\frac{v2 - v1}{l} + \sin\alpha\right)\frac{1}{l\beta^3}\left(\frac{v2 - v1}{l} + \sin\alpha\right)\frac{-1}{l}$$

To assemble all of the elements, the total energy has the following scalar form

$$\sum_e \{\delta U^{(e)}\}^T \{KU^{(e)}\} = \{\delta U\}^T \{KU\} \tag{4.13}$$

where $U = X$ is the global nodal displacement vector and

$$\{Ku\} = \{g\} = \begin{bmatrix} g_3^1 + g_3^2 + g_1^4 + g_1^5 \\ g_4^1 + g_4^2 + g_2^4 + g_2^5 \\ g_3^3 + g_3^4 + g_1^6 + g_1^7 \\ g_4^3 + g_4^4 + g_2^6 + g_2^7 \\ g_3^5 + g_3^6 + g_1^8 + g_1^9 \\ g_4^5 + g_4^6 + g_2^8 + g_2^9 \\ g_3^7 + g_3^8 + g_1^{10} + g_1^{11} \\ g_4^7 + g_4^8 + g_2^{10} + g_2^{11} \\ g_3^9 + g_3^{10} + g_1^{12} + g_1^{13} \\ g_4^9 + g_4^{10} + g_2^{12} + g_2^{13} \\ g_3^{11} + g_3^{12} + g_1^{14} + g_1^{15} \\ g_4^{11} + g_4^{12} + g_2^{14} + g_2^{15} \\ g_3^{13} + g_3^{14} + g_1^{16} \\ g_4^{13} + g_4^{14} + g_2^{16} \\ g_3^{15} + g_3^{16} \\ g_4^{15} + g_4^{16} \end{bmatrix} \tag{4.14}$$

Therefore we have a dynamic equation of motion for the nonlinear truss shown in Figures 6 and 7. The nonlinear differential equation of motion is

$$M\ddot{X} + D\dot{X} + \{KU\} = F(t) \tag{4.15}$$

where M is the mass matrix, D is the damping matrix, and $\{KU\}$ is a nonlinear stiffness term. This dynamic equation is a fully nonlinear model for the truss. The following sections will analyze the static solution, structural optimization, and active vibration-suppression for the nonlinear truss.

2.4.4.1. Characteristics of the Nonlinear Truss Model

The static and dynamic characteristics of the nonlinear truss are examined before the truss design is optimized and controlled. The physical constants E, A, and l are assumed respectively as follows: $E = 10.6 \times 10^6$ lb/inch^2 (identical for all 16 elements); $A = 1/3$ inch^2 (identical for all 16 elements, constrained by $0.3 \leq A \leq 0.45$ for the optimization); $l = 12$ in for elements 1, 3, 5, 7, 9, 11, 13, 15; $l = 8$ in for elements 4, 8, 12, 16; and $l = 14.4222$ in for elements 2, 6, 10, 14. Thus the overall truss is 48 inches long and 8 inches high. A lumped mass matrix is given below.

$$\mathbf{M} = \text{diag}[5.6\,5.6\,5.6\,5.6\,5.6\,5.6\,5.6\,5.6\,5.6\,5.6\,5.6\,5.6\,5.5\,5.5\,5.3\,5.3]*10/16$$

The mass matrix contains about 90% non-structural mass at each node of the truss. The nonconservative energy is due the external load at the end of the truss, and damping is assumed proportional to the linear stiffness matrix ($\mathbf{D} = 0.0002\mathbf{K}_{\text{linear}}$) of the truss.[54]

Nonlinear structures can have very complicated dynamic behavior. To help understand the behavior of the nonlinear truss model, the static solution is considered as shown,

$$\{\mathbf{KU}\} = \mathbf{F} \tag{4.16}$$

where $\mathbf{F}$ is a constant external force vector. By varying the external force, we can characterize the relationship between the external force and the displacement of the truss. The Optimization Toolbox of MATLAB provides several ways to solve nonlinear equations represented by $\mathbf{f}(\mathbf{x}) = 0$ where $\mathbf{f}(\mathbf{x})$ is a nonlinear equation. The MATLAB function used here is CONSTR for constrained minimization. The resulting displacement at DOF 14 is shown in Figure 8 with respect to the external force $F(t)$ shown in Figure 6. From Figure 8 it is obvious that the nonlinear truss model has the characteristic of a hard spring. The deflected shape of the truss for a large load is shown in Figure 9.

The displacement responses of the nonlinear truss under the external forces of $F(t) = 3000 \sin(\omega(t)/\sqrt{2}$, $F(t) = 30000 \sin(\omega t)/\sqrt{2}$, and $F(t) = 50000 \sin(\omega t)/\sqrt{2}$ where $\omega = 4\pi$ are given in Figures 10 to 12. The notation underneath the figure titles specifies the load case (F_m), gain coefficients (CT), and the value of the objective function (J) which includes three terms corresponding to displacement $\mathbf{x}$, velocity $\dot{\mathbf{x}}$, and control force $\mathbf{u}$. The multiple curves show the nodal responses of the truss, which have increasing amplitude corresponding to the increased distance of the nodes from the support.

The uncontrolled truss exhibits a beating response that occurs because the forcing frequency of $f = 2$ Hz is very close to the first natural frequency of a linear model of the truss. At the larger loads the truss response jumps to a large nonzero equilibrium point and vibrates about this point with a reduced amplitude. This response is due to the nonlinear stiffness of the truss but physically would not happen due to the interference or buckling of the truss members. Note that the x (odd DOFs) response is one-sided (i.e. asymmetric) for the higher load cases. This is because the cantilever truss is stiff in extension but flexible in bending, and the load direction causes bending

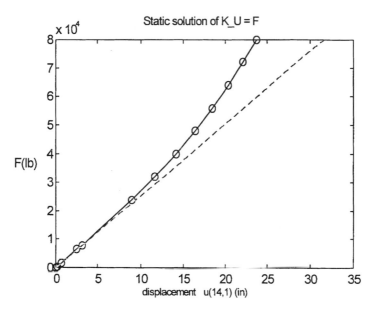

Figure 8. Static deflection of nonlinear truss due to tip load.

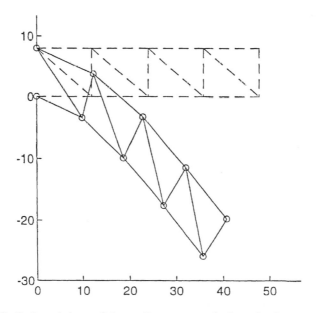

Figure 9. Deflected shape of the nonlinear truss under large load.

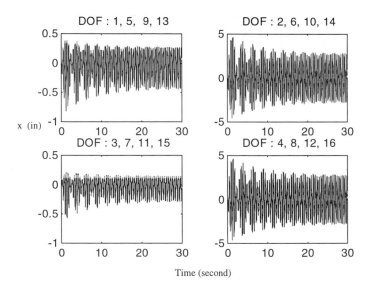

Figure 10. Fully nonlinear truss responses for $F_m = 3000$ without feedback $CT = [0\ 0\ 0]$; $J = 13.5671 = 13.3478 + 0.2193 + 0$.

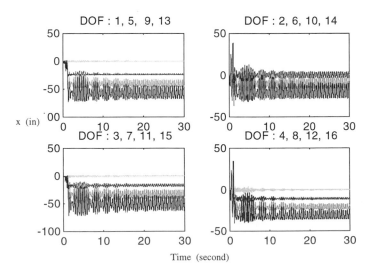

Figure 11. Fully nonlinear truss Responses for $F_m = 30000$ without feedback $CT = [0\ 0\ 0]$; $J = 11457 = 11443 + 14 + 0$.

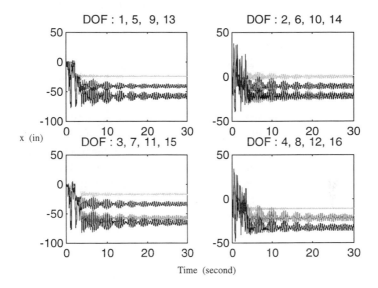

Figure 12. Fully nonlinear truss responses for F_m = 50000 without feedback $CT = [0\ 0\ 0]$; $J = 19389 = 19359 + 30 + 0$.

and the negative displacement in the x direction. The objective function values for these three load cases without control are listed in Table 2. These loads and resulting displacements are large compared to the size of the truss in order to exhibit the nonlinear behavior. The nonlinear behavior would occur at lower load levels for a longer or more flexible truss.

As a check on the accuracy of the technique, the nonlinear design algorithm was used to calculate gains for different static loads with the actuator located at the load. The resulting gain coefficients for a cubic polynomial control law are $c1 = 2528.7$, $c2 = 6.3$, and $c3 = 1$ for the linear, quadratic and cubic terms, respectively, and these coefficients fit the static load curve in Figure 8. This checks that the control algorithm and optimizer are working correctly, and indicates that the linear term will be dominant in the control law. This curve fitting function can be used to decide a rough form for a control law for a nonlinear structure, even though the static analysis does not reflect the dynamic characteristics completely.

2.4.4.2. Active controller design and optimization of structural parameters

The general control approach taken here is to consider any type of nonlinear control function that uses constant gain output feedback control. Feedback

Table 2. Objective function J values for different load cases without feedback (cross-section area = 1/3, response time = 30 sec)

Load $F_m =$	3,000 (lb)	30,000 (lb)	50,000 (lb)
$J =$	13.5671	11457	19389

control using constant gains is much simpler than using a dynamic compensator, and output feedback is appropriate for large structures where only a small number of sensors are practical. Stability is in general not guaranteed when using nonlinear control, but stability robustness may be improved by using collocated active damping and collocated position-feedback. Another approach being investigated is based on the idea that a simple spring-damper system added between two coordinates on the structure will not destabilize the system. This nonlinear feedback control law uses the absolute value of a polynomial function multiplied by the sign of the velocity or displacement at the actuator location (called a discontinuous control law), or polynomial functions (called a continuous control law). This means that the control forces act as if they were produced from a spring-damper system with time-varying coefficients. The stability and robustness of this type of control law is being investigated.

Integration parameters corresponding to constant acceleration over each time step ($\beta = 0.25, \lambda = 0.5$) are used in the integration and initial conditions are zero. A cubic polynomial control law is used with feedback of the velocity and position of only one point at the end of the truss ($x_{16} = y_8$). The general control law is shown below,

$$[\mathbf{u}_v]_2 = c_{1v}{}^*\dot{x}_{16} + c_{2v}{}^*\dot{x}_{16}{}^*|\dot{x}_{16}| + c_{3v}{}^*\dot{x}_{16}^3$$
$$[\mathbf{u}_p]_2 = c_{1p}{}^*\dot{x}_{16} + c_{2p}{}^*\dot{x}_{16}{}^*|\dot{x}_{16}| + c_{3p}{}^*\dot{x}_{16}^3 \tag{4.17}$$

where $\mathbf{u}_p$ is the position feedback control force, $\in \mathbf{R}^{16\times 1}$, and $[\mathbf{u}_p]_j$ is the jth element in $\mathbf{u}_p$, and means that an active controller is installed at the jth DOF, $\mathbf{u}_v$ is the velocity feedback control force, $\in \mathbf{R}^{16\times 1}$, and the c_{iv}, c_{ip} values are coefficients to be determined by the optimization. The weighting matrices chosen are $\mathbf{Q}_p = \mathbf{I}, \mathbf{Q}_v = 10^{-5}\mathbf{I}$, and $\mathbf{R}_p = 10^{-12}\mathbf{I}, \mathbf{R}_v = 10^{-12}\mathbf{I}$, where $\mathbf{I}$ is the identity matrix. This particular weighting puts almost all emphasis on minimizing the displacements of the truss with little penalty for control force magnitude. The nonlinear design technique is used to design an active controller and to simultaneously optimize the structural parameters to suppress the vibration of the truss. The optimization duration is 2.5 seconds. Through computer simulation, it was found that a stepsize of 0.01 second is too large and makes the integration unstable. A step size of 0.001 seconds is satisfactory for small loads and for large loads with small errors in the gradient, but a timestep of 0.0001 seconds is necessary for the largest load to obtain a nearly exact gradient calculation, as compared to the MATLAB finite-difference gradient calculation. The specific equations are derived in this section to optimize the cross-sectional area of the elements in the truss and to design non-collocated and collocated, linear and nonlinear position feedback controllers. Damping design will be discussed in section 4.5. Other forms of continuous control laws not discussed can also be designed using the nonlinear design technique.

2.4.4.3. Continuous position-feedback controller design

A polynomial control law is investigated here. Damping design is much more complicated than position feedback control law design, and for clarity only position feedback control is considered in this section. Let

$$\mathbf{u}_v(\dot{\mathbf{x}}_{r+1}) = \mathbf{0}_{16 \times 1}$$

$$[\mathbf{u}_p(\mathbf{x}_{r+1})]_j = \xi_2 x_m + \xi_3 x_m |x_m| + \xi_4 x_m^3 \qquad (4.18)$$

$$\xi = [\xi_1 \ \xi_2 \ \xi_3 \ \xi_4]$$

where the design variable ξ_1 is the cross-section area, and ξ_2, ξ_3 and ξ_4 are the controller gains corresponding to the c values in Equation (4.17), and if $j = m$, $\mathbf{u}_p$ is a collocated controller, otherwise it is a non-collocated controller. Therefore the following specific derivatives with respect to the design variables are:

$$\frac{\partial \mathbf{u}_v(\dot{\mathbf{x}}_{r+1})}{\partial \xi} = \mathbf{0}_{16 \times 4} \qquad (4.19)$$

$$\frac{\partial [\mathbf{u}_p(\mathbf{x}_{r+1})]_j}{\partial \xi} = [0 \ x_m \ xm|x_m| \ |x_m^3]$$

$$+(\xi_2 + 2|x_m|\xi_3 + 3\xi_4 x_m^2) \left[\frac{\partial x_m}{\partial \xi_1} \ \frac{\partial x_m}{\partial \xi_2} \ \frac{\partial x_m}{\partial \xi_3} \ \frac{\partial x_m}{\partial \xi_4} \right]_j \qquad (4.20)$$

where j indicates the jth row in the matrix. The derivative of the local stiffness term with respect to ξ_1 is the following

$$\frac{\{\mathbf{KU}^{(e)}\}}{\partial \xi_1} = \begin{bmatrix} El\varepsilon_{11} \dfrac{\partial \varepsilon_{11}}{\partial u_1} \\[2mm] El\varepsilon_{11} \dfrac{\partial \varepsilon_{11}}{\partial u_2} \\[2mm] El\varepsilon_{11} \dfrac{\partial \varepsilon_{11}}{\partial u_3} \\[2mm] El\varepsilon_{11} \dfrac{\partial \varepsilon_{11}}{\partial u_4} \end{bmatrix} + \frac{\partial \{\mathbf{KU}^{(e)}\}}{\partial \mathbf{X}} \frac{\partial \mathbf{X}}{\partial \xi_1} \qquad (4.21)$$

where

$$\frac{\partial \{\mathbf{KU}^{(e)}\}}{\partial \mathbf{X}} = \frac{\partial}{\partial \mathbf{X}} \begin{bmatrix} EAl\varepsilon_{11}\phi_1 \\ EAl\varepsilon_{11}\phi_2 \\ EAl\varepsilon_{11}\phi_3 \\ EAl\varepsilon_{11}\phi_4 \end{bmatrix} = EAl \begin{bmatrix} \phi_1\frac{\partial}{\partial \mathbf{X}}\varepsilon_{11} + \varepsilon_{11}\frac{\partial}{\partial \mathbf{X}}\phi_1 \\ \phi_2\frac{\partial}{\partial \mathbf{X}}\varepsilon_{11} + \varepsilon_{11}\frac{\partial}{\partial \mathbf{X}}\phi_2 \\ \phi_3\frac{\partial}{\partial \mathbf{X}}\varepsilon_{11} + \varepsilon_{11}\frac{\partial}{\partial \mathbf{X}}\phi_3 \\ \phi_4\frac{\partial}{\partial \mathbf{X}}\varepsilon_{11} + \varepsilon_{11}\frac{\partial}{\partial \mathbf{X}}\phi_4 \end{bmatrix}$$

For the optimization process, the procedure to calculate the gradient is given in section 4.3. First we calculate $\frac{\partial \ddot{\mathbf{x}}_{r+1}}{\partial \xi}$, then $\frac{\partial \mathbf{x}_{r+1}}{\partial \xi}$, and finally the closed-form gradient of the objective function $\frac{\partial J}{\partial \xi_j}$. The algorithm CONSTR in MATLAB is used to optimize all of the design variables. Both a linear and a nonlinear control law can be designed in a similar way by using the method mentioned above.

From Figure 8 it is obvious that the nonlinearity increases with increasing external force $F(t)$. To show this effect, we will consider active control cases using the external forces defined for the open-loop case. We will use the amplitude of $Fm = 3000, 30000$ and 50000 to denote three different load cases, although the actual loads are divided by $\sqrt{2}$. The design variables are subject to constraints and have initial values of $\xi = \mathbf{0}_{4 \times 1}$ for the optimization. If a linear control law is designed, everything is the same as in nonlinear case except $\xi_3 = 0$ and $\xi_4 = 0$. For the position-feedback controller we only consider the noncollocated controller, which is more practical than a collocated one. Here the output feedback measurement signal is only the displacement of DOF 16, as shown in Figure 6. The actuator is installed between DOF 2 and ground.

Two types of noncollocated controllers are designed. The first is linear for the different F_m respectively. The design results for the linear noncollocated control are the following: $lp3 : \xi_2 = 47209; lp30 : \xi_2 = 46000;$ $lp50 : \xi_2 = 21019$; where $lp3, lp30$ and $lp50$ represent the linear position feedback constants obtained corresponding to the three load cases for F_m. Later np, lv and nv similarly represent feedback constants for nonlinear position feedback, linear velocity feedback (damping), and nonlinear velocity feedback (damping), respectively. Note that the optimization for the $lp50$ case did not reach the global optimum. The design results for the nonlinear noncollocated control are the following: $np3 : \xi_{2,3,4} = [40511 \; 568.7 \; 8.73]; np30 : \xi_{2,3,4} = [37148 \; 8.591 \; 1.2229]; np50 : \xi_{2,3,4} = [30751 \; 2.6233 \; 0.5015]$. The nonlinear cases are not exactly at the global minimum when the corresponding linear cases give slightly lower J values. After the optimization, ξ_1 always reached its upper bound of 0.45 in^2. The responses and control force for the $np50$ case are shown in Figures 13 and 14, respectively. The beating and bifurcation occurring in the uncontrolled truss are eliminated and the vibration amplitude is greatly reduced. To further reduce the vibration at the end of the truss

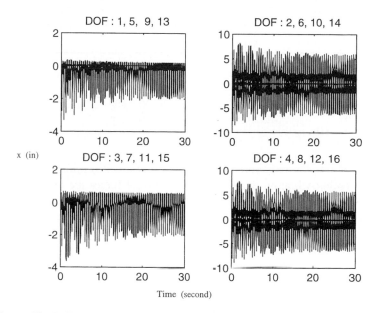

Figure 13. Fully nonlinear truss responses for $F_m = 50000$ with noncollocated $np50$ $CT = 1.0e + 004 * [3.4871 \ 0.6424 \ 0.1337]$; $J = 54.9936 = 53.8749 + 1.0955 + 0.0232$.

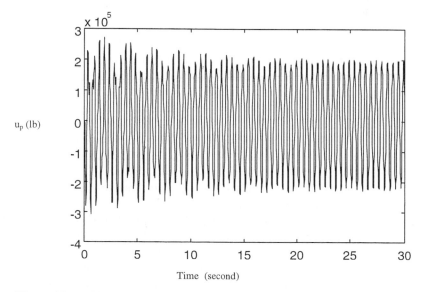

Figure 14. $np50$ position-feedback force for $F_m = 50000$.

Table 3. Objective function J values for different load cases with non-collocated position feedback ($lp3/30/50$ represents control laws $lp3$, $lp30$, or $lp50$ applied to load case F_m)

	$lp3/30/50$	$np3/30/50$	$np50$
$F_m = 3000$	0.1259/$lp3$	0.1520/$np3$	0.1869/$np50$
$F_m = 30000$	13.586/$lp30$	16.224/$np30$	19.079/$np50$
$F_m = 50000$	106.17/$lp50$	54.994/$np50$	54.994/$np50$

would increase the vibration at the actuator and this is a limiting condition in suppressing vibration of the truss. The control forces are large because the actuator force is applied near the support where the truss is stiff, while the excitation is at the flexible end of the truss, and because there is a large moment arm effect between the control force and load. This example shows that active control and the effect of increasing the cross-sectional area of the truss members reduces the peak vibration amplitude of the truss by about 80 percent. The corresponding objective function values are listed in Table 3. Also given in Table 3 are J values when the $np50$ control law is applied to all load cases. These results indicate that in nonlinear design there are different solutions to the control law that give similar performance, and that the optimization solution should be run longer ($>$ 2.5 seconds of response) for a final design to give more precise results. The nonlinear control law, $np50$, designed for the maximum load condition was also shown to be stable for all other lower load cases.

The data in Table 3 shows that with large control forces, linear and nonlinear control laws have about the same performance for this type of structure subjected to a harmonic excitation when using an objective function that weights the response much greater than control force. In general the nonlinear control can operate over a wider range of load conditions and can be designed to "turn off" for low disturbances to reduce control power consumption. Different types of loading (e.g. random or transient), structural nonlinearities, actuator locations, and control weighting parameters are all expected to give somewhat different conclusions. This points out that there is a large amount of research that still needs to be done to make nonlinear control practical for many applications.

Overall, from the simulations performed[54], for large deflections nonlinear control has a small advantage over linear control. From a design standpoint, nonlinear control is also better because it helps the optimization converge to near the global minimum. A universal nonlinear control law can be designed by assuming the maximum load on the structure. While low load is suitable for the design of a linear control law, the maximum load should always be used for the design of a nonlinear control law because the nonlinear system may become unstable if the load is increased above the maximum design point. A comparison of responses before and after the optimization are shown in Figures 15 and 16, where the $lp50$ control law is applied to the $F_m = 3000$ and $F_m = 30000$ load cases, respectively. This shows that active control can prevent catastrophic nonlinear effects from occurring for extreme load cases.

This section has shown that the fully nonlinear truss model reflects the physical characteristics of very flexible structures and that nonlinear position feedback has a small advantage over linear control to suppress vibration in large structures, and helps the optimization converge. When the control forces are large, vibrations are brought down to the linear range and the linear controller is as effective as the nonlinear controller. The general design steps for a large scale flexible structure are: (1) Build a nonlinear model of the structure; (2) Apply a maximum load; and (3) Use the nonlinear design technique to optimize the design of the structure, and to design a linear or nonlinear position feedback control law. The optimization solutions for this example took from 4 to 20 hours wall time to complete using a 133 MHz PC, depending on the timestep used. Also, each run was usually repeated to ensure that the global optimum solution was found. This points out that nonlinear design is very time consuming and computationally intensive, and that parallel computing will be needed to speed-up computations to solve large nonlinear structures problems.

2.4.4.4. *Discontinuous position-feedback controller design*

Four types of discontinuous controllers were studied for active position-feedback control or damping control. They are:

$$(1) \quad u_p = \text{sgn}(x_{ic}) |c_1 x_m + c_2 x_m |x_m| + c_3 x_m^3| \qquad (4.22a)$$

$$(2) \quad u_p = x_{ic} |c_0 + c_1 x_m + c_2 x_m |x_m| + c_3 x_m^3| \qquad (4.22b)$$

$$(3) \quad u_p = \text{sgn}(x_{ic}) |c_1 x_m + c_2 |x_m|^{\frac{1}{2}} \text{sgn}(x_m) + c_3 x_m^{\frac{1}{3}}| \qquad (4.22c)$$

$$(4) \quad u_p = x_{ic} |c_0 + c_1 x_m + c_2 |x_m|^{\frac{1}{2}} \text{sgn}(x_m) + c_3 x_m^{\frac{1}{3}}| \qquad (4.22d)$$

where x_{ic} is the signal (measured if necessary) at the actuator's DOF, and x_m is the measurement signal from the sensor at DOF m. When an absolute function $|f(x)|$ is considered, we have the first derivative of $|f(x)|$ with respect to the variable x as the following

$$\frac{\partial |f(x)|}{\partial x} = \begin{cases} \frac{\partial f(x)}{\partial x} & f(x) \geq 0 \\ \frac{\partial f(x)}{\partial x} & f(x) < 0 \end{cases}$$

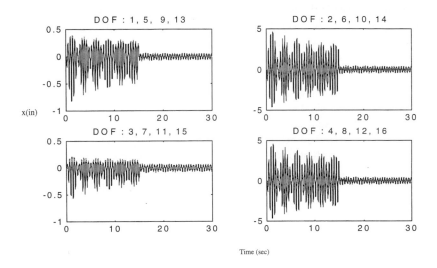

Figure 15. Open ($A = 1/3$) and closed ($A = 0.45$) loop responses of the fully nonlinear truss for $F_m = 3000$ with $lp50$.

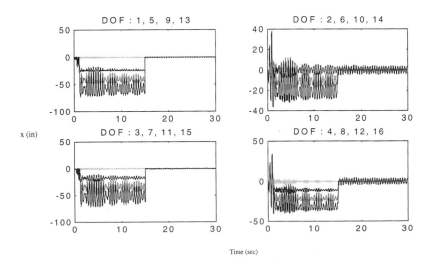

Figure 16. Open ($A = 1/3$) and closed ($A = 0.45$) loop responses of the fully nonlinear truss for $F_m = 30000$ with $lp50$.

If $f(x)$ is a continuous function, the gradient equation above is sufficiently accurate for numerical calculation (it should be noted that $f(x)$ will never be zero in numerical calculation by computer). Unfortunately, the $u_p(x)$ considered in Equations (4.22) is a discontinuous function, which means that $|u_p(x)|$ has some jump points. Therefore the above gradient Equation is incorrect, and the MATLAB finite-difference gradient is also incorrect at jump points. Thus it is difficult to use a discontinuous feedback controller with the nonlinear design technique because the objective function becomes discontinuous and the optimization diverges. A smooth approximation of the sign function is given in[5], and this may allow the discontinuous control law to work, but this has not been investigated yet.

2.4.5. Collocated Damping Feedback Controller Design

In general, damping is a very effective way to suppress vibration. It is especially useful in lightweight structures applications where other methods such as a vibration absorber may be impractical because an added mass is needed. In the following we consider active damping for the fully nonlinear truss, including non-collocated and collocated nonlinear and linear damping. In the last section, we considered noncollocated position-feedback because it is more effective than collocated feedback for engineering applications. Conversely, it will be shown that collocated damping is more practical for this example.

It is assumed that the signal used for damping is from the DOF at which the damper is installed (i.e. collocated active damping). From the previous section, it was concluded that the nonlinear control law (using polynomial terms) is better than the linear control law if the displacement is large and the nonlinearity is obvious. Here nonlinear damping will be tested, using the same formulation as the nonlinear position-feedback law. The nonlinear damping control law is:

$$u_v = c_1 \dot{x}_m + c_2 \dot{x}_m |\dot{x}_m| + c_3 \dot{x}_m^3 \qquad (4.23)$$

For the nonlinear truss discussed, the active damper is located at the measurement DOF of $m = 6$ and the same external forces will be considered. The linear damping control law is

$$u_v = c_1 \dot{x}_m \qquad (4.24)$$

During the optimization design, the elemental cross-section area is considered as a design variable. Results for the damping design optimization are given as follows: Gain constants for collocated linear velocity damping are: $lv3 : c_1 = 6345.9$; $lv30 : c_1 = 6046.6$; and $lv50 : c_1 = 5981.6$; and gain constants for

Table 4. Objective function J values for different load cases with collocated active damping at DOF 6

	$lv3/30/50$	$nv3/30/50$	$nv50$
$F_m = 3000$	$0.1369/lv3$	$0.1480/nv3$	$0.1365/nv50$
$F_m = 30000$	$14.014/lv30$	$13.312/nv30$	$13.560/nv50$
$F_m = 50000$	$38.840/lv50$	$37.254/nv50$	$37.254/nv50$

collocated nonlinear velocity damping are: $nv3 : c = [5381.1\ 43.51\ 379.13]$; $nv30 : c = [6608.2\ 0.2134\ 0.09699]$; $nv50 : c = [6378.0\ 1.3001\ 0.33442]$. Responses and corresponding control forces for the $nv50$ case are shown in Figures 17 and 18. The objective function values are listed in Table 4. These results show a slight advantage for nonlinear collocated damping as compared to linear collocated damping. The case in Table 4 where nonlinear damping is worse than linear damping ($J = 2.0268$ versus $J = 1.8831$) is because the optimization did not find the global minimum. Nonlinear damping design should always work as well or better than linear damping at the same load point because linear damping is just a special case of the nonlinear damping control law. It should also be mentioned that for a non-harmonic forcing function, nonlinear damping may have a more significant advantage.

2.4.5.1. *Non-collocated damping feedback controller design*

In engineering problems, we often consider getting a measurement signal from the most sensitive position and to put actuators at the most convenient position, especially for large structures. That means non-collocation between sensors and actuators. In the last section, it is shown that a non-collocated displacement controller can achieve good performance in vibration suppression. However, for the case of damping, further investigation is needed. Unfortunately it is difficult or almost impossible to design noncollocated damping for this example. This is because nonlinear damping control causes high frequency control signals to be fed back to the actuator, and this leads to unstable responses. Velocity responses for one case are shown in Figure 19, before they become unstable during the optimization process. Discontinuous damping similar to the position feedback laws shown in equations (4.22) was also tried, but the gradient calculation becomes incorrect and the optimization cannot converge. As mentioned, smoothing the gradient at jump points is an area for further research. The stability of nonlinear controllers is discussed in more detail in the next section.

Conclusions for the damping cases are similar to those for the position-feedback cases in last section, in particular, that nonlinear damping has a small

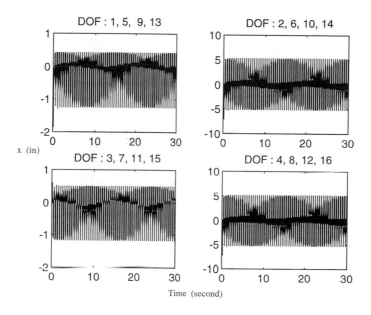

Figure 17. Fully Nonlinear Truss Responses for $F_m = 50000$ with collocated $nv50$ $CT = 1.0e + 003 * [6.3780 \ 1.3001 \ 3.3442]$; $J = 37.2542 = 36.6567 + 0.5947 + 0.0028$.

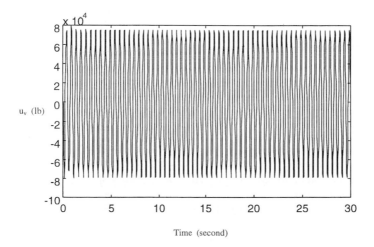

Figure 18. $nv50$ damping force for $F_m = 50000$.

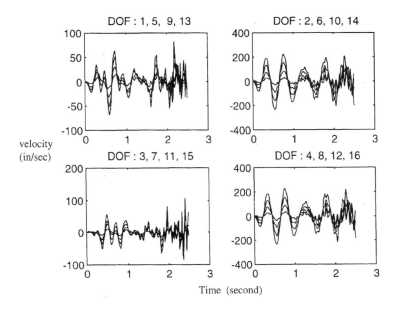

Figure 19. Velocity responses of unstable optimization process with noncollocated damping feedback (measurement DOF is 16 and actuator DOF is 2).

advantage over linear damping for the example considered. Non-collocated damping did not work because the phase of the velocity signal at the measurement point changes relative to the velocity at the actuation point and causes the close-loop system to become unstable.

2.4.6. Stability of Nonlinear Feedback Controllers

Nonlinear non-collocated damping is difficult to design because the feedback signal contains higher frequency components. These higher frequencies are introduced into the truss in the nonlinear damping control and lead to response instability. The opposite result was attained for non-collocated position-feedback, which works well. These results are explained using a simple linear model.

Consider a single-degree-of-freedom linear model

$$a\ddot{x} + b\dot{x} + cx = F(t) - e\dot{y} \qquad (4.25)$$

where e is a gain constant, $\dot{y}$ is a velocity measurement, and $a, b, c, e > 0$.

If $\dot{x}$ isn't equal to zero, Equation (4.25) can be rearranged as

$$a\ddot{x} + (b + e\dot{y}/\dot{x})\dot{x} + cx = F(t) \tag{4.26}$$

There are two eigenvalues which define the stability of the solution,

$$r_{1,2} = \frac{-(b + e\dot{y}/\dot{x}) \pm \sqrt{(b + e\dot{y}/\dot{x}^2 - 4ac}}{2a} \tag{4.27}$$

If either of the real parts of $r_{1,2}$ are positive, the solution becomes unstable. Thus if

$$\text{Re}(-(b + e\dot{y}/\dot{x}) + \sqrt{(b_e\dot{y}/\dot{x})^2 - 4ac}) > 0, \quad \text{that is if } (b + e\dot{y}/\dot{x}) < 0, \quad \text{or}$$

$$\frac{\dot{y}}{\dot{x}} < -\frac{b}{e}, \tag{2.38}$$

then the solution is unstable. For the truss example considered here (also generally for structural control), b represents structural damping and is usually very small. Therefore $(b + e\dot{y}/\dot{x})$ can easily become negative if $\dot{x}$ and $\dot{y}$ become out of phase. That is why noncollocated nonlinear damping is very difficult to design, and the success of using velocity feedback for nonlinear systems is problematic and configuration dependent.

Similarly we can consider an active position feedback controller using a simple model

$$a\ddot{x} + b\dot{x} + cx = F(t) - dy \tag{4.29}$$

where d is a gain constant, and y is a position feedback measurement. It is obvious that if $d > 0$ and x and y are collocated, the solution of Equation (4.29) will be stable. But if $d > 0$, and x and y are non-collocated, we have the following equation

$$a\ddot{x} + b\dot{x} + (c + dy/x)x = F(t) \tag{4.30}$$

which has two eigenvalue roots

$$r_{1,2} = \frac{-b \pm \sqrt{b^2 - 4a(c + dy/x)}}{2a} \tag{4.31}$$

If y and x are non-collocated, then they may become out of phase. The solution of Equation (4.31) will become unstable if
$\text{Re}(-b + \sqrt{b^2 - 4a(c + dy/x)}) > 0$, where $(a, b, c, d > 0)$, that is, if
$4a(c + dy/x) < 0$, or

$$\frac{y}{x} < -\frac{c}{d}. \tag{4.32}$$

Assuming that x is not equal to 0, then if y/x satisfies the last equation, the noncollocated position feedback will be unstable. Through comparison of the damping case and the position feedback case, the condition for stability for position feedback is easier to achieve because typically the ratio c/d for position feedback would be larger than the ratio b/d for velocity feedback. Also, the sign and absolute value of the ratio of the measurement and actuation responses is more constant for displacements (y/x) than for velocities $(\dot{y}/\dot{x})$ for a noncollocated nonlinear system. This is shown in Figure 19 where higher frequency components due to velocity feedback are making the response go unstable.

This simple single degree-of-freedom model for stability gives insight to why it is possible to control one mode of a system using a noncollocated actuator and sensor, but difficult to control many modes without using full state feedback or collocation between actuators and sensors.

2.4.7. Linear System Design using the Time-Integration Algorithm

The nonlinear design technique developed above can also be used to design linear structures and controllers. One approach developed is to perform a modified modal reduction by separating the structural matrices into a baseline structure $(\mathbf{M}, \mathbf{K})$ and update matrices $(\mathbf{M}u, \mathbf{K}u)$. The modal reduction is done using the modes from the baseline structure and in this case the update matrices are not diagonalized by the modal transformation. However, the model size is reduced and the parameters in the update matrices can be simultaneousely optimized with the control gain matrices in modal coordinates using the nonlinear design algorithm. The modified modal reduction follows the transformation:

$$\mathbf{x} = \mathbf{PH} \tag{4.33}$$

$$\mathbf{P}^T \mathbf{MP} = \mathbf{I} \tag{4.34}$$

$$\mathbf{P}^T \mathbf{KP} = \mathrm{diag}(\omega_i^2) \tag{4.35}$$

$$\mathbf{h} + \mathbf{P}^T \mathbf{M}_u \mathbf{Ph} + \mathrm{diag}(2\zeta_i \omega_i)\mathbf{h} + \mathrm{diag}(\omega_i^2)\mathbf{h} + \mathbf{P}^T \mathbf{K}_u \mathbf{Ph} = \mathbf{P}^T \mathbf{f} - \mathbf{P}^T \mathbf{u} \tag{4.36}$$

where $\mathbf{h} = [\eta_1 \ \eta_2 \ldots \ \eta_m]^T$ is the $m \times 1$ vector of modal coordinates, $\mathbf{P} = [\mathbf{p}_1 \ \mathbf{p}_2 \ \ldots \ \mathbf{p}_m]$ is an nxm matrix of mode shapes $\mathbf{p}_i$ where the m mode shapes correspond to the frequency range of interest, $\omega_i^2 (i = 1, 2, \ldots, m)$ are the squared natural frequencies, $\mathbf{f}$ is the external force vector, ζ_i are the modal damping ratios, and the control forces $\mathbf{u}$ depend on $\mathbf{x}$ (or $\dot{\mathbf{x}}$), and also η (or $\dot{\eta}$) in modal coordinates. The modal reduction is recomputed after a fixed number of iterations of the optimization so that accuracy is maintained. This approach reduces computations from using a full size structural model and allows a larger time step to be used in the integration solution, and has been successfully applied to the design of a vibration suppression system using a piezoceramic inertial actuator. A challenging extension would be to apply the modified modal approach to nonlinear structures using nonlinear normal modes[48].

2.4.8. Summary of Nonlinear Design Technique

The Computational Nonlinear Design technique presented can simultaneously optimize structures and controllers by directly using the second-order nonlinear or linear equations of motion. In nonlinear design, one typically has to find equilibrium points and linearize the model before a damper or a controller can be designed. For the design of large structures, this would be difficult to do because of model size and complex nonlinearities, and because the structural characteristics are changing during the design optimization. The proposed method avoids this pre-analysis, and integrates the structure and controller design problems into a computationally efficient procedure.

To test the technique, the design of a geometrically exact nonlinear truss was optimized. This included the design of different types of control laws, and simultaneous optimization of the elemental cross-sectional area. It was shown that noncollocated linear and nonlinear position feedback control laws, and collocated damping feedback control laws worked well, and that nonlinear control has a small advantage over linear control for the example considered. The method is not yet applicable to discontinuous control laws. The technique can also optimize any parameter in the structure, including Youngs Modulus $E^{(e)}$, structural shape parameters such as elemental length $l^{(e)}$, elemental cross-section area $A^{(e)}$, and elemental angle $\alpha^{(e)}$. Structural shape optimization using the nonlinear design technique is more accurate than linear design because the geometrically exact deformed shape of the structure is computed.

Ways to improve the technique include improving the robustness of the control with respect to the variations in excitation forces and structural and gain parameters, and testing a continuous gradient function to allow discontinuous control laws to be used. Different types of controllers such as fuzzy logic and neural network controllers could be studied for comparison to the polynomial type currently used. Other types of nonlinear finite elements should be added to the code, and approximate closed-form gradients derived. Parallel-vector sparse matrix equation-solvers and optimization techniques are also needed to apply the present technique to large scale structures. This would be an important step toward developing a New Generation of Nonlinear Structural Analysis Tools for Shared and Distributed Memory High-Performance Computers.

2.5. CONCLUSIONS

Although there are many accepted techniques to design controllers and to optimize structures separately, there is no universal technique to simultaneously design structures and structural controllers. This is because optimal control algorithms are difficult to couple to a structural optimization, and adjusting the controller and structural variables simultaneousely using multivariable optimization is computationally very intensive and the global optimum solution is difficult to find when adjusting a large number of design variables. In spite of the difficulties, multi-disciplinary design optimization is becoming necessary to use to meet increasingly stringent design requirements, and more research effort is needed to develop practical techniques that can integrate structure, controller and overall system designs, and to use multiple processors to allow larger models to be solved. Techniques for simultaneous nonlinear design and control for highly flexible structures are just beginning to be developed and nonlinear dynamics research needs to be extended from one and two DOF models to large practical systems to define the advantages of nonlinear design.

The three integrated design techniques presented in detail in this chapter are based on developing as small of a structural model as possible, and using the full model in the structural optimization and controller design. This approach to simultaneous design optimization improves accuracy over reduced basis solution techniques, and takes advantage of the second order form of smaller size matrices that are sparse and symmetric to reduce computations, and can use parallel-vector computing to solve large problems.

Acknowledgement

The authors thank the following researchers for their contributions to the material in this chapter; Dr. P. Frank Pai derived the nonlinear truss model used in Section 4 and his suggestions improved all of the techniques in this Chapter, Mr. Xubin Song developed the nonlinear design algorithm and performed the analysis in Section 4, and Mr. Sunil Thyagarajan performed the analysis in Section 2.

References

1. Kamat, M.P., 1993, *Structural Optimization: Status and Promise*, AIAA, Washington, DC.
2. Haug, E.J. and J.S. Arora, 1979, *Applied Optimal Design*, Wiley-Interscience, New York.
3. Khot, N.S. and R.T. Haftka, 1993, *Structures And Controls Optimization*, ASME, New York.
4. R.V. Grandhi, G. Bharatram, and V.B. Venkayya, 1993, *AIAA Journal*, **31**(7), 1329–1337.
5. Junkins, J.L. and Y. Kim, 1993, *Introduction to Dynamics and Control of Flexible Structures*, AIAA, Washington, DC.
6. Junkins, J.L., 1990, *Mechanics and Control of Large Flexible Structures*, AIAA, Washington, DC.
7. Soong, T.T., 1990, *Active Structural Control, Theory and Practice*, Longman Scientific, New York.
8. Kohudic, M.A., 1994, *Advances in Vibration Control for Intelligent Structures*, Technomic Publishing, Basel, Switzerland.
9. Khot, N.S. and S.A. Heise, 1994, *AIAA Journal*, **32**(3), 610–615.
10. Khot, N.S., F.E. Eastep, H. Oz and V.B. Venkayya, 1986, *AIAA Paper 86-0842*, pp. 43–53.
11. Kajiwara, I., K. Tsujioka, and A. Nagamatsu, 1994, *AIAA Journal*, **32**(4), 866–873.
12. Schulz, M.J. and D.J. Inman, 1995, *Journal of Sound and Vibration*, **182**(2), 259–282.
13. Schulz, M.J. and D.J. Inman, 1994, *IEEE Journal on Control Systems Technology*, **2**(2), 88–100.
14. Waters, D.P. and M.J. Balas, 1996, *AIAA-96-1227-CP*, pp. 223–230.
15. Vivek Mukhopadhyay, 1995, *Journal of Aircraft*, **32**(1), 45–51.
16. Adams, M.M.Jr. and D.M. Christhilf, 1995, *Journal of Aircraft*, **32**(1), 52–60.
17. Woods-Vedeler, J.A., A.S. Potozky, and S.T. Hoadley, 1995, *Journal of Aircraft*, **32**(1), 68–76.
18. Waszak, M.R. and S. Srinathkumar, 1995, *Journal of Aircraft*, **32**(1), 61–67.
19. Fuessel, D., T. Singh, and C.L. Bloebaum, 1995, *AIAA-95-1480-CP*, 2964–2972.
20. Groumpos, P.P., 1994, *IEE Proc.-Control Theory Appl.*, **141**(1), 1–11.
21. Bucher, I. and S. Braun, 1994, *Journal of Sound and Vibration*, **175**(4), 433–453.
22. Bucher, I. and S. Braun, 1994, *Journal of Sound and Vibration*, **175**(4), 455–473.
23. Balas, G.J. and P.M. Young, 1995, *Journal of Guidance, Control and Dynamics*, **18**(2), 325–332.
24. Beyers, R. and S. Desa, 1994, *Journal of Mechanical Design*, **116**(2), 396–404.
25. Khot, N.S., F.E. Eastep, H. Oz, and V.B. Venkayya, 1986, *AIAA Paper 86-0842*, pp. 43–53.
26. Utku, S. and B.K. Wada, 1994, *NASA TECH BRIEF*, **18**, 11, Item #29, JPL New Technology Report NPO-18820, Pasadena, California.
27. Wang, D.S., G. Yang, and M. Donath, 1993, *Proceedings of the American Control Conference*, San Francisco, CA, pp. 552–559.
28. Burdisso, R.A. and C.R. Fuller, 1994, *Journal of Guidance, Control and Danamics*, **17**(3), 466–472.
29. Hyland, D.C., 1990, *Proceedings of the 29th Conference on Decision and Control*, Honolulu, Hawaii.
30. Hall, E.K.II and S.V. Hanagud, 1993, *Journal of Guidance, Control and Dynamics*, **16**(3), 470–476.
31. Shiau, T. and A. Jean, 1990, *Journal of Vibration and Acoustics*, **112**, 501–507.
32. Friedmann, P.P. and T.A. Millott, 1995, *Journal of Guidance, Control and Dynamics*, **18**(4), 664–673.
33. Thyagarajan, S.K., M.J. Schulz, and J.C. Slater, 1994, *First University/Industry Symposium On High Speed Civil Transport Vehicles*, N.C.A&T State University, pp. 408–413.
34. Schulz, M.J., S.K. Thyagarajan, and J.C. Slater, 1995, *AIAA Journal*, **33**(8), 1486–1491.
35. Thyagarajan, S.K., 1995, M.Sc. Thesis, *Frequency Response Function Optimization for Mechanical Systems*, Department of Mechanical Engineering, North Carolina, AT&T State Univeristy.
36. MATLAB USERS GUIDE, 1995, *The Math Works*, Natick, MA.

37. Jin, I.M. and L.A. Schmit, 1992, *AIAA Journal*, **30**(7), 1892–1900.
38. Grace, A., 1995, *Optimization Toolbox for Use With MATLAB, Users Guide*, The Math Works, Natick, MA.
39. Szyszkowski, W. and J.M. King, 1993, *AIAA Journal*, **31**(11), 2163–2168.
40. Levy, R., 1993, *NASA Tech. Brief*, 17, 9, Item #71, JPL New Technology Report, NPO-18774, Pasadena, California.
41. Smith, M.J. and S.G. Hutton, 1992, *AIAA Journal*, **30**(7), 1886–1891.
42. Grandhi, R., 1993, *AIAA Journal*, **31**(12), 2296–2303.
43. Visser, W.J and M. Imregun, 1991, *Proceedings of the 9th International Modal Analysis Conference*, Florence, Italy, pp. 462–468.
44. Venkayya, V.B., V.A. Tischler, and N.S. Knot, Dynamics and Control of Space Structures, (AFWAL/FIBRA, Design and Analysis Methods Group, Analysis and Optimization Branch, Structures and Dynamics Division, WPAFB, Dayton, Ohio).
45. Slotine, J. and W. Li, 1991, *Applied Nonlinear Control*, Prentice-Hall.
46. Sardar, H.M. and M. Ahmadian, 1992, *Journal of Vibration and Acoustics*, **114**, 154–160.
47. Goh, C.J., 1994, *International Journal of Control*, **114**(1), 91–115.
48. Slater, J.C., 1993, *Nonlinear Modal Control*, Ph.D. Thesis, State University of New York at Buffalo, New York.
49. Taylor, J.H. and J. Lu, 1993, *Proceedings of the American Control Conference*, San Francisco, CA, pp. 536–540.
50. Isidori, A. and W. Kang, 1995, *IEEE Transactions on Automatic Control*, **40**(3), 466–472.
51. James, M.R. and J.S. Baras, 1995, *IEEE Transactions on Automatic Control*, **40**(6), 1007–1017.
52. Adiguzel and H. Oz, 1991, *AIAA-91-2808, Guidance, Navigation and Control Conference*, New Orleans, LA, pp. 1–9.
53. Song, M.J. Schulz, and P.F. Pai, 1996, *Sixth Conference on Nonlinear Vibrations, stability, and Dynamics of Structures*, Virginia Polytechnic Institute and State University, Blacksburg, VA, pp. 1–2.
54. Song, 1996, *Design and Control of Nonlinear Structures*, MS Thesis, Department of Mechanical Engineering, North Carolina A&T State University, Greensboro, NC.
55. Agarwal, T.K., O.O. Storaasli, and D.T. Nguyen, 1990, *AIAA Paper 90-1149*, Presented at the AIAA/ASME/ASCE/AHS 31st Structures, Structural Dynamics and Materials Conference, Long Beach California, pp. 1-11.
56. Baddourah, M.A., O.O. Storaasli, E.A. Carmona, and D.T. Nguyen, 1991, *AIAA-91-1006-CP*, pp. 1547–1553.
57. Nguyen, D.T., O.O. Storaasli, J. Qin, and R. Qamar, 1993, *Multidisciplinary Parallel-Vector Computation Center*, 135 KDH Building, Old Dominion University, Norfolk, VA 23529.
58. Nguyen, D.T., O.O. Storaasli, E.A. Carmona, A. Nasra, Y. Zhang, M.H. Baddourah, and T.K. Agarwal, 1991, *Computing Systems in Engineering*, **2**(3).
59. Belvin, W.K., P.G. Maghami, and D.T. Nguyen, 1992, *Computing Systems in Engineering Journal*, **3**(1–4), 181–188.
60. Pai, P.F. and A.H. Nayfeh, 1994, *Computers & Structures*, **53**, 877–895.
61. Pai, P.F. and A.H. Nayfeh, 1994, *Int. J. Solids and Structures*, **31**, 1309–1340.

3 NONLINEAR MODAL CONTROL TECHNIQUES AND APPLICATIONS IN STRUCTURAL DYNAMIC SYSTEMS

J.C. SLATER[1] and G. AGNES[2]

[1]*Department of Mechanical and Materials Engineering, Wright State University, Dayton, OH 45435*
[2]*Department of Engineering Science and Mechanics, Virginia Polytechnic Institute and State University, Blacksburg, VA 24060*

3.1. PREFACE

The control of multiple degree of freedom nonlinear systems is perhaps the most challenging control problem engineers face today. Nonlinear oscillatory systems are quite common (i.e. robotics, slewing, large strain vibrations, vibration of bimodulus composite materials, the motion of a swinging spring, the vibrations of shells and composite plates...) and the nonlinear control of these systems is becoming more possible with the advent of ever faster microprocessors.[17] Popular methods being used today include Lyapunov design, adaptive control, linearizing control, sliding mode control, fuzzy logic and neural networks.[31] Lyapunov based control design can be unwieldy for nonlinear systems with many degrees of freedom. Linearizing control is likely to be wasteful of energy, since the linearizing part of the control design does nothing to improves actual performance. Often it is not feasible since the actuators and sensors must be placed in locations such that they are able to counteract the nonlinear effects. This is usually possible only for nonlinear effects in actuators. Fuzzy logic and neural network controls assume an ambiguity that does not necessarily exist, and thus can add unwarranted complexity to the control problem. One concept which has not been fully exploited with respect to nonlinear systems is that of nonlinear modal control.

The concept of nonlinear modal control is to apply the Invariant Manifold method of Shaw and Pierre[2][5] for finding nonlinear modes and extend the method to include forced excitation and output equations. Once the dynamic equations of motion have been transformed to the nonlinear modal coordinate system, it is possible to extend the concept of modal norms for linear systems to the case of systems with nonlinear modes. The approximately "decoupled" nonlinear modal equations also allow the design of control laws for each nonlinear single degree of freedom system using standard methods (such as Lyapunov design for each individual modal equation). Once control laws have been designed for the single degree of freedom modal systems, the control is transformed back into the original state space for controlling the original system. It is demonstrated in examples that nonlinear modal control can be used to satisfy multiple control objectives which linear modal control alone cannot.

3.2. NOTATION

q, r element of the specified matrix corresponding to the qth row and the rth column

q, R element of the specified matrix corresponding to the qth row and the Rth columns

even   even numbered rows of a non-square matrix or even numbered rows and columns of a square matrix.

$\tilde{A}(z)$ state functions

A, C usual state space $2N \times 2N$ matrices

$\tilde{A}_i(z)$ i^{th} element of $\tilde{A}(Z)$

$A(z)$: $\tilde{A}(z) = A(z)z$

$\tilde{A}_m(w)$:   modal state functions

$\tilde{A}_{m_i}(w)$ i^{th} element of $\tilde{A}_m(w)$

$A_m(w)$: $\tilde{A}_m(w) = A_m(w)w$

$A_{md}(w)$:   desired modal state matrix

$A_{md}(w)$:   desired modal state functions

α, β constants representing the linear relationship between x_i and y_i

B usual state space $2N \times N$ matrices

B_f $N \times N$ force matrix for the system in linear 2nd order form

B_{fn} normalized B_f

B_{mn} B_m normalized

B_{mpn} B_{mp} normalized

B_{mp}, B_m modal forcing matrix

B_{mp_q} the qth row of the matrix B_{mp}

C_p output matrix for states

C_{pm} modal output matrix for states

C_{pmn} normalized modal output matrix for states

C_v output matrix for the derivatives of the states

$\tilde{\mathbf{C}}(\mathbf{z}, \hat{\mathbf{u}}_z)$ observation function in physical coordinates

$\tilde{\mathbf{C}}_m(\mathbf{z}, \hat{\mathbf{u}}_z)$ observation function in modal coordinates

$\mathbb{C}_{q,x}$ controllability norm of the qth mode from the rth actuator

$f(t)$ arbitrary forcing function

Φ phase angle

$G_z(\mathbf{z})$: modal control law function matrix

$G_m(\mathbf{w})$: modal control law

G_p: control feedback matrix for state vector

G_v: control feedback matrix for derivative of state vector

I identity matrix

I_q identity matrix size N_q

Λ eigenvalues of state matrix A

$\mathbf{m}$ nonlinear mode vector

$_i\mathbf{m}$ nonlinear mode vector of ith mode

$M(\mathbf{w})$: $\mathbf{z} = M(\mathbf{w})\mathbf{w}$

$\tilde{\mathbf{M}}(\mathbf{w})$: $\mathbf{z} = \tilde{\mathbf{M}}(\mathbf{w})$

M, D, K mass, damping and stiffness matrices

$N(\mathbf{z})$: $\mathbf{w} = N(\mathbf{z})\mathbf{z}$

$\tilde{\mathbf{N}}(\mathbf{z})$: $\mathbf{w} = \tilde{\mathbf{N}}(\mathbf{z})$

N_Q number of distinct natural frequencies

N_q multiplicity of the qth natural frequency

$\phi_{q,r}$ observability norm of the qth mode from the rth sensor

ω natural frequency

Ω diagonal matrix of natural frequencies

Ω_g diagonal matrix of the natural frequency of the qth repeated mode

P_{fp} diagonal matrix of sensor sensitivities

P_{fp} decoupling matrix for modally damped systems

P_{snp} matrix of associated eigenvectors of state matrix A

P_{fnp} eigenvectors of A_m

P_s eigenvectors $M^{-1/2} K M^{-1/2}$

P_{um} transforms response into modal coordinates

P_n normalization matrix

$S(\mathbf{w}, \hat{\mathbf{u}}_z)$ sensitivity of $\mathbf{y}$ to $\mathbf{w}$

u, v modal coordinates

$\hat{\mathbf{u}}$ force vector in physical coordinates

$\hat{u}_i$ i^{th} element of $\hat{\mathbf{u}}$

$u_{\hat{x}}, u_{\hat{y}}, u_{\hat{z}}$ deflections in the $\hat{x}, \hat{y}, \hat{z}$ directions

$\hat{\mathbf{u}}_m$:   force vector in modal coordinates

$\hat{\mathbf{u}}_z$ force vector in linear second order form

$\mathbf{w}$:   modal coordinate space

x_i, y_i displacement and velocity of the ith degree of freedom, respectively

x_c, y_c chosen displacement and velocity pair

$\hat{x}, \hat{y}, \hat{z}$ location of a point on a structure in three dimensional space

$X_i(u, v), Y_i(u, v)$ functions which relate the displacements x_i, y_i to the modal coordinates u and v

V_i potential energy of the ith spring

$\hat{\mathbf{y}}$ output variable

$\hat{\mathbf{y}}_n$ normalized output variable

$\mathbf{z}$:   physical coordinate space

Z diagonal matrix of damping ratios

Z_q diagonal matrix of the damping ratio of the qth repeated mode

3.3. INTRODUCTION

Since the advent of the computer, ever more complex and sophisticated linear control methods have been, and still are being, developed for vibration suppression. One of the first and simplest is modal control. The concept of linear modal control is to decouple the systems equations of motion, choose a control law for each mode individually, and transform the modal control law from modal coordinates to physical coordinates in order to implement it on the structure. This works well in cases where it is feasible. Unfortunately, the conditions which must be met are quite strict, and usually cannot be met. Although the modal control can often be approximately applied even when these conditions are not met, other advanced techniques have surpassed modal control in practical application (pole placement, H_∞, H_2, LQR, LQG, HAC/LAC, and others). However, such techniques are not applicable to systems with significant nonlinearities. Nonlinear modal control is the first nonlinearly analogous application of a linear control.

The limitations of these control methodologies is that they assume a linear model for the structure, either by neglecting the nonlinear terms, approximating the nonlinear terms with linear terms, or applying linearizing control in order to make the structure act as if it were linear. Neglecting the nonlinearities in the control design can cause closed loop instabilities or poor performance. Linearizing control performed to simplify the system dynamics in order to allow the application of linear control techniques is usually only practical for linearizing actuator and sensor dynamics. The linearizing control effort does not necessarily improve the system response. In some cases, nonlinearities can often improve the system response.

All structures and systems that oscillate are nonlinear. When a structure is referred to as linear, what is meant is that the effects of the nonlinearities are so small that a linear model is sufficient. Nonlinearities which are often neglected, justifiably so in some cases, are air damping, boundary condition damping, boundary condition stiffness effects, geometrical effects (where the equations of motion are written in terms of the undeformed state), centripetal forces, coriolis forces, and other miscellaneous nonlinearities which are often negligible within the accuracy of measurement or modeling capabilities. However, nonlinear effects are often far from negligible and readily observed in certain systems. Some examples are the vibration of beams, strings, plates, membranes and shells for which stretching is significant; the motion of structures with nonlinear springs; the motion of pendulums; the motion of a rocking ship; and slewing motions in which the centripetal forces are significant[7]. These structures exhibit what are called *nonlinear normal modes*. The discovery of nonlinear normal modes is generally credited to Rosenberg[24]. This work was expanded by Atkinson and Taskett[2], Szempliska[32], Rand[20,21,22], Greenburg and Yang[8], Yen[40], and others. Vakakis[33] summarizes much of this work in his dissertation. More recent work in nonlinear normal modes has been by King and Vakakis[13], Nayfeh and Nayfeh[18], and Shaw and Pierre[25,26] and Shaw[27]. The nonlinear modal control methods demonstrated in this work are based on extending the Invariant Manifold technique[25] to active control applications.

In this work, we consider the forced equations of motion including the output equation. The modal norms proposed by Hamden and Nayfeh[9] and Hughes and Skelton[10] are expanded to account for variation of sensor sensitivity and actuator strength. Applying the nonlinear modal coordinate transformations derived by Shaw and Pierre[25] allows the derivation of nonlinear modal state equations including the output equation from which observability and nonlinear modal controllability norms are proposed. In systems with high nonlinearity, linear norms can be misleadingly small since the nonlinearities are not accounted for. The norms derived here include the amplitude dependence of the ability to sense and actuate. Using these transformations, it is also demonstrated that control can be derived for nonlinear modal systems in the modal space and subsequently transformed into physical coordinates for implementation.

Three examples, one linear and two nonlinear, demonstrate the method. The first example demonstrates the concepts developed in this work applied to a linear system. The second example demonstrates design of a control law for a nonlinear system with similar normal modes such that global performance characteristics can be achieved from the controlled modal system. No linearization or linearizing control is used. The third example demonstrates the implementation of a control law, designed in modal space, in physical coordinates. The control law is nonlinear in the modal coordinates as well as the physical coordinate system. Simulations of the nonlinear modally controlled system are compared to simulations of the system with a corresponding linear control. The simulations show that the desired control effects are accomplished with the nonlinear modal control, thus showing that nonlinear modal control is a viable alternative to other control methodologies.

3.4. LINEAR MODAL CONTROL

The most common method of control law design is to ignore nonlinearities, assuming their effects are negligible relative to the linear approximation, and design a control law based on the linear model. The second most common is to apply linearizing control to the system such that the system acts as if it is linear. Then a control law is designed for the control linearized system. A third method not described here is to apply robust control, such as H_∞, including the nonlinearities as uncertainties. Although this will lead to robust stability, it will not provide optimal performance. Clearly the best control law should result from a design process using all of the available model information. Linear modal control is discussed as one of the simplest ways to achieve vibration suppression. Linear observability and controllability norms are discussed and expanded to account for the simultaneous use of different actuators and sensors. Controllability and observability norms are also derived in terms of the state variables in order to more accurately determine optimal sensor and actuator placement.

3.4.1   Linearization of a Nonlinear System

A common approach for controlling systems is to linearize the weakly nonlinear equations of motion such that linear control may be applied. The most common method is to take a Taylor series of the equations of motion about the operating point and drop the nonlinear terms. Although numerous other methods have been devised for finding "equivalent" linear systems, Taylor series expansion is the most common because it is the simplest to apply and can be applied to systems with large numbers of degrees of freedom. The state equations of motion for a nonlinear system can be written in the form

$$\dot{\mathbf{z}} = \bar{\mathbf{A}}(\bar{\mathbf{z}}) + B(\bar{\mathbf{z}})\hat{\mathbf{u}}_z \tag{1}$$

First, the equilibrium values of the states ($\bar{\mathbf{z}}_0$) are found by taking the mean value $\hat{\mathbf{u}}_{z_0}$ of the expected excitation $\hat{\mathbf{u}}_z$ and setting $\dot{\mathbf{z}} = \mathbf{0}$ in Equation (1) such that

$$\mathbf{0} = \tilde{\mathbf{A}}(\mathbf{z}_0) + B(\mathbf{z}_0)\hat{\mathbf{u}}_{z_0} \tag{2}$$

Once $\bar{\mathbf{z}}_0$ and $\hat{\mathbf{u}}_{z_0}$ have been found, $\tilde{\mathbf{A}}(\bar{\mathbf{z}})$ and $B(\bar{\mathbf{z}})\hat{\mathbf{u}}_z$ are expanded in a multivariable Taylor series about the equilibrium such that

$$\begin{aligned}
\dot{\mathbf{z}} =& \tilde{\mathbf{A}}(\mathbf{z}) + B(\bar{\mathbf{z}}_0)\hat{\mathbf{u}}_{z_0} \\
=& \bar{\mathbf{A}}(\bar{\mathbf{z}}_0) + B(\bar{\mathbf{z}}_0)\hat{\mathbf{u}}_{z_0} + \left.\frac{\partial\tilde{\mathbf{A}}(\mathbf{z})}{\partial\bar{\mathbf{z}}}\right|_{\bar{\mathbf{z}}_0}(\bar{\mathbf{z}} - \bar{\mathbf{z}}_0) + \left.\frac{\partial B(\bar{\mathbf{z}})\hat{\mathbf{u}}_z}{\partial\bar{\mathbf{z}}}\right|_{\bar{\mathbf{z}}_0,\hat{\mathbf{u}}_{z_0}} \\
& (\mathbf{z} - \mathbf{z}_0) + B(\mathbf{z}_0)(\hat{\mathbf{u}}_z - \hat{\mathbf{u}}_{z_0})
\end{aligned} \tag{3}$$

where the partial derivatives of a vector with respect to a vector represents the Jacobian. The linear state equations can then be written in the form

$$\dot{\mathbf{z}} = A\mathbf{z} + B\mathbf{u} \tag{4}$$

where

$$\mathbf{z} = (\bar{\mathbf{z}} - \bar{\mathbf{z}}_0), \qquad \mathbf{u} = (\hat{\mathbf{u}}_z - \hat{\mathbf{u}}_{z_0}) \tag{5}$$

the elements a_{ij} of the matrix A are

$$a_{ij} \left.\left(\frac{\partial\bar{\mathbf{A}}_i}{\partial\bar{\mathbf{z}}_j}\right)\right|_{\bar{z}=\bar{z}_0} + \left.\left(\frac{(\partial B\hat{\mathbf{u}}_z)_i}{\partial\bar{\mathbf{z}}_j}\right)\right|_{\bar{\mathbf{z}}=\bar{\mathbf{z}}_0,\hat{\mathbf{u}}_z=0} \tag{6}$$

and the elements b_{ij} of the matrix B are the elements b_{ij} of $B(\bar{\mathbf{z}}_0)$.

Numerous reasons can exist for not linearizing the model. Often, the nonlinear terms can dominate the system dynamics or important parts of the dynamics. For example, a lightly damped system oscillating in a fluid will be subject to damping proportional to the velocity squared. When linearized using this method, the fluid damping is neglected since there is no linear term in the Taylor series expansion. Methods do exist, for example Inman[12], for finding a non-zero equivalent damping term, but in linearizing the damping, the amplitude dependence is still ignored in the final linear model. Linearizing a system containing cubic stiffness terms ignores one major benefit of having a cubic stiffness. When a system with a softening spring is excited sinusoidaly, the frequency of excitation can sweep through resonance in the positive direction (increasing frequency) without a large response. The converse is true as well. A number of detrimental effects can be overlooked when nonlinearities are ignored. "von Kármán observed that certain parts of an airplane can be violently excited by an engine running at an angular speed much larger than their natural frequencies, and Lefschetz described a commercial aircraft in which the propellers induced a subharmonic vibration in the wings which in turn induced a subharmonic vibration in the rudder. The oscillations were violent enough to cause tragic consequences."[17] Another observation of nonlinear systems is that of Nayfeh, Mook and Marshall[16] where certain systems excited at a natural frequency ω_1 where a natural frequency $\omega_2 = 2\omega_1$ exists can exhibit no steady state response even in the presence of damping.

Although numerous methods exist for finding linearized models[14,4,15], the end result is always a set of equations in which potentially important information has been ignored.

3.5. LINEAR MODAL CONTROL, OBSERVABILITY, AND CONTROLLABILITY

3.5.1. Modal Decoupling of a Linear System

In a modally damped linear system, a modal motion is defined by the linear functional interrelationship between the displacements of the system's degrees of freedom. For a non-modally damped linear system, this functional interrelationship involves the displacements and velocities of the system's degrees of freedom. What this means is that for a modally damped system, when the motion takes place in a single mode, if the displacement of any point is known than the displacements of all other degrees of freedom can be found as a linear function of that displacement. Also, if the velocity of any point is known than the velocities of all other degrees of freedom can be found as a linear function of that velocity. For a non-modally damped system moving in a single mode, if the displacement and velocity of any point is known, then the displacements and velocities of all other degrees of freedom are a linear function of that displacement and velocity. That is, the displacement and velocity of each degree of freedom may be represented as a linear combination of the displacement and velocity of a chosen degree of freedom:

$$x_i = \alpha_{1i} x_c = \alpha_{2i} y_c \qquad (7)$$

$$y_i = \beta_{1i} x_c = \beta_{2i} y_c \qquad (8)$$

Here x_i and y_i represent the displacement and velocity of the ith degree of freedom, x_c and y_c represent the chosen displacement/velocity pair, and α and β are constants representing the linear relationship between the displacements and velocities. In the case of modal damping, the displacement of each degree of freedom may be represented a constant times the displacement of a chosen degree of freedom and the velocity of each degree of freedom may be represented by a constant times the velocity of the chosen degree of freedom, i.e. $\alpha_{2i} - \beta_{1i} = 0$. For a system with N degrees of freedom, this creates two vectors of numbers (length $N - 1$) relating the relative displacements and velocities of all of the degrees of freedom.

The relationship between the modal coordinate and the mode shape can be chosen by simultaneously scaling the modal vector and modal coordinate. Thus the relationship between the modal coordinates (velocity and displacement) and any physical coordinate can be defined by an arbitrary constant times a linear combination of the displacement and velocity of the chosen coordinate. For linear systems, a mode can be normalized such that this constant is 1 for the chosen coordinate. Thus the chosen coordinate displacement and velocity are equal to the modal coordinate displacement and velocity. This corresponds to the chosen element of the mode shape vector being set equal to one when using eigenanalysis to find the modal characteristics of a system.

For a linear system, it is possible for the relationship between the mode shape vector and the modal coordinate to be chosen to be nonlinear. Although there is little use for this, it is instrumental as a demonstration of what the modal relationship represents. Take for instance a two degree of freedom for which one of the modal equations is

$$\ddot{u} + u = f(t) \tag{9}$$

where u is the modal coordinate and $f(t)$ is an arbitrary forcing function. Also, take the relationships between the modal coordinate u and the two displacements x_1 and x_2 to be

$$x_1 = u, \tag{10}$$

and

$$x_2 = 2u. \tag{11}$$

From Equations (10) and (11), the mode shape can be represented as $x_2 = 2x_1$. Using Equations (9), (10) and (11) the motion of each of the degrees of freedom can be determined when the motion of the system takes place only in this mode. This is not, however, a unique way of writing the modal relations. Other relations exist which will yield the same physical motions when moving in this mode but yield a different modal equation.

For example, we can also represent the modal motion of the system by the following set of equations. The modal equation of motion is written as

$$\frac{d^2u^3}{dt^2} + u^3 = 6u\dot{u}^2 + 3u^2\ddot{u} + u^3 = f(t) \tag{12}$$

The displacement of the first degree of freedom is

$$x_1 = u^3 \tag{13}$$

and the displacement of the second degree of freedom is

$$x_1 = 2u^3 \tag{13}$$

Substituting Equations (13) and (14) into Equation (12) and substituting Equations (11) and (10) into (9), one can verify that the motion of the displacements x_1 and x_2 for the second system are identical to the motion of the displacements x_1 and x_2 for the first system. Although this analysis may not be useful for linear systems, when nonlinear modal systems are considered, an understanding of these concepts facilitate the understanding of why multiple representations of the same mode exist.

For proportionally damped linear second order systems, modal decomposition can be performed in second order form. However, for non-modally damped systems, the system must be written in first order state space form to perform modal decomposition. Since the proportionally damped system can be put into the more general first order state space form as well, we will discuss modal control of systems in state space form. The first order state space representation of a linear second order matrix form system can be written as

$$\dot{\mathbf{z}} = A\mathbf{z} + B\mathbf{u}_z$$
$$\mathbf{y} = C_p\mathbf{z} + C_v\dot{\mathbf{z}} \tag{15}$$

where

$$\begin{bmatrix} 0 & I \\ -M^{-1} & -M^{-1}D \end{bmatrix} \tag{16}$$

and

$$\begin{bmatrix} 0 \\ B_f \end{bmatrix} \tag{17}$$

Here A, C_p, and C_v represent the usual $2N \times 2N$ (N is the number of degrees of freedom of the system) second order linear matrices, B represents the usual $2N \times N$ forcing matrix, B_f represents the $N \times N$ force matrix for the system in linear second order form and $\bar{\mathbf{u}}_z$ represents the force vector in linear second order form. For modally damped systems, the first order state equations may be decoupled by substituting $\mathbf{z} = P_{fp}\mathbf{w}$ into (15) and premultiplying P_{fp}^{-1} by where

$$\begin{bmatrix} P_s & 0 \\ 0 & P_s \end{bmatrix} \tag{18}$$

and

$$P_s = \mathrm{eig}(M^{-1}K) \tag{19}$$

The modal state space equations may then be written as

$$\dot{\mathbf{w}} = \begin{bmatrix} 0 & I \\ -\Omega^2 & -2Z\Omega \end{bmatrix} \mathbf{w} + \begin{bmatrix} 0 \\ B_{mp} \end{bmatrix} \mathbf{u}_z = \begin{bmatrix} 0 & I \\ -\Omega^2 & -2Z\Omega \end{bmatrix} \mathbf{w} + \mathbf{u}_m \tag{20}$$

$$\mathbf{y} = C_p P_{fp} \mathbf{w} + c_v P_{fp} \dot{\mathbf{w}}$$

where

$$B_{mp} = P_s^{-1} B_f \tag{21}$$

is the modal forcing matrix. Here the vector $\mathbf{w}$ represents the state space modal coordinate vector $[\mathbf{x}^T \ \dot{\mathbf{x}}^T]^T$ where $\mathbf{x}$ is the vector of modal displacements and $\mathbf{u}_m = [\mathbf{0} \ B_{mp}^T]^T \mathbf{u}_z$ is the modal force vector.

The unforced non-modally damped system can be transformed into modal coordinates by finding the diagonal matrix of eigenvalues Λ and matrix of associated eigenvectors P_{snp} of the state matrix A. The following relations can then be used to construct the modal state matrix in the same form as Equation (20):

$$\Omega = \mathrm{abs}(\Lambda) \qquad Z = -(\mathrm{Re}(\Lambda))^{-1}\Omega \tag{22}$$

Using eigenvalue/eigenvector decomposition on the state matrix A and the modal state matrix A_m, the matrix P_{fnp} can be found such that

$$P_{fnp}^{-1} A P_{fnp} = A_m \tag{23}$$

Unfortunately, for the non-modally damped system (a system with complex modes), it is not possible to maintain the state forcing matrix form of the proportionally damped system. The form of the state equations for the non-modally damped system is then

$$\dot{\mathbf{w}} = \begin{bmatrix} 0 & I \\ -\Omega^2 & -2Z\Omega \end{bmatrix} \mathbf{w} + \begin{bmatrix} \Sigma \\ B_{mnp} \end{bmatrix} \mathbf{u}_z$$

$$\mathbf{y} = C_p P_{fnp} \mathbf{w} + c_v P_{fnp} \dot{\mathbf{w}} \tag{24}$$

Note the matrix Σ which causes the violation of the definition of the state space modal coordinate vector by modifying the first N equations. These first N equations represented identity equations for the modally damped and unforced non-modally damped systems. In modifying these equations by adding a force, the concept of modal control is violated since it is not possible to control a mode without changing the mode shape. Thus for the forced non-modally damped system, the concept of a mode is lost. Papers to date on modal control of linear systems have failed to confront the problem of complex modes well. Two fairly well known papers on modal controllability and observability of linear second order systems are by those by Hamden and Nayfeh[9], and Hughes and Skelton[10]. The two works are very similar in that the method of Hamdan and Nayfeh[9] is a geometrical interpretation of the method of Hughes and Skelton[10]. The method of Hughes and Skelton decouples the system and then applies controllability and observability conditions to determine controllability and observability norms. By bypassing the decoupling of the system, Hamdan and Nayfeh ignore the matrix Σ which can effect the controllability of a non-modally damped system and thus derive erroneous results for non-modally damped systems. Because of this effect, Hughes and Skelton have dealt primarily with undamped or proportionally damped systems. Since this work is an extension of this theory to nonlinear normal modal systems, all systems considered here will be modally damped.

3.5.2. Modal Excitation of a Linear System

For a linear system, finding the response to a modal excitation requires transforming the force into modal coordinates as well as transforming the response into modal coordinates. Define the vector $\mathbf{u}_z$ to be

$$\mathbf{u}_z = P_{um}\mathbf{u}_m \qquad (25)$$

where $\mathbf{u}_m$ is a $N \times 1$ vector of desired modal forces. If a matrix P_{um} exists such that

$$B_{mp}\mathbf{u}_z = B_{mp}P_{um}\mathbf{u}_m = I\mathbf{u}_m \qquad (26)$$

where I is the identity matrix then each mode can be forced individually without affecting the other modes. Thus the condition

$$P_{um} = B_{mp}^{-1} \qquad (27)$$

must hold and B_{mp} must be nonsingular. Since

$$B_{mp} = P_s^{-1}B_f \qquad (28)$$

and P_s is nonsingular, the necessary and sufficient condition for B_{mp} to be non-singular is that B_f be nonsingular. A more lenient but almost as rare condition is that if a matrix P_{um} can be found such that in any single column of $B_{mp}P_{um}$ there exists a non-zero element in the ith position and all other elements of the given column are zero then the ith mode is individually controllable. In other words, when an attempt is made to excite one mode by itself, other modes are excited as well unless this condition is met. (The exception being the steady state sinusoidal excitation of a system at a natural frequency with no undamped modes.) However, it is still possible to observe the modal forces and the system response to those forces.

The forced response of the system of Equation (15) is

$$\mathbf{z}(t) = e^{At}\mathbf{z}(0) + \int_o^t e^{A(t-\tau)} B\mathbf{u}_z(t)d\tau \tag{29}$$

Since

$$\mathbf{z} = P_{fp}\mathbf{w} \tag{30}$$

the modal response to a force $\mathbf{u}_z$

$$\mathbf{w}(t) = P_{fp}^{-1} e^{At} P_{fp}\mathbf{w}(t)\Big|_{t=0} + P_{fp}^{-1} \int_0^t e^{A(t-\tau)} B\mathbf{u}_z(t)d\tau \tag{31}$$

and the modal response to modal excitations is

$$\mathbf{w}(t) = P_{fp}^{-1} e^{At} P_{fp}\mathbf{w}(t)\Big|_{t=0} + P_{fp}^{-1} \int_0^t e^{A(t-\tau)} B P_{um}\mathbf{u}_m(t)d\tau \tag{32}$$

3.5.3. Modal Controllability of a Linear System

Regardless of whether the ith mode is individually controllable, the ability to *affect* a mode of interest is the principle concern when attempting to perform control. The degree to which one can affect a mode is also of interest since optimal placement and actuator sizing can make the control problem easier and improve the controlled results. In addition, a low controllability of a higher frequency mode can prevent unintended spillover of control effort into the mode.

For the system written in the form of Equation (15), the necessary and sufficient condition for controllability of the system is[19]

$$\text{rank}[B \ AB \ A^2B \ \cdots \ A^{2N-1}B] = 2N \tag{33}$$

For vibration suppression, a more important way to consider the controllability of each mode. A combined understanding of the system's natural frequencies, the expected excitation forces, and the controllability norms are extremely useful in the determination of proper actuator placement.

Consider the system of Equation (15) with modal damping. The modal form of the state equation is

$$\dot{\mathbf{w}} = \begin{bmatrix} 0 & I \\ -\Omega^2 & -2Z\Omega \end{bmatrix} \mathbf{w} + \begin{bmatrix} 0 \\ B_{mp} \end{bmatrix} \mathbf{u}_z \tag{34}$$

Since the system is now decoupled, it can be broken down into N_Q independent systems where N_Q is the number of distinct natural frequencies. The equation of motion for the qth system is

$$\dot{\mathbf{w}}_q = \begin{bmatrix} 0 & I \\ -\Omega_q^2 & -2Z_q\Omega_q \end{bmatrix} \mathbf{w}_q + \begin{bmatrix} 0 \\ B_{mp_q} \end{bmatrix} \mathbf{u}_z \tag{35}$$

where Iq is an identity matrix of size Nq and Nq is the multiplicity of the qth natural frequency. The matrix Ω_q is the $N_q \times N_q$ diagonal matrix of the qth natural frequency and B_{mp_q} are the corresponding rows of the matrix B_{mp}. The controllability condition for the qth system becomes (see Equation (33))

$$\text{rank} \begin{bmatrix} 0 & B_{mp_q} & 0 & \cdots & (-\Omega_q^2)^{N_q-1} B_{mp_q} \\ B_{mp_q} & 0 & -\Omega_q^2 B_{mp_q} & \cdots & 0 \end{bmatrix} = 2N_q \tag{36}$$

which reduces to

$$\text{rank}[B_{mp_q} \quad -\Omega_q^2 B_{mp_q} \quad \cdots \quad (-\Omega_q^2)^{N_q-1} B_{mp_q}] = N_q \tag{37}$$

As shown by Hughes and Skelton[10] this is a special case of the Jordan-Form Controllability theorem derived in Chen.[6] Equation (37) then reduces to

$$\text{rank}[B_{mp_q}] = N_q \tag{38}$$

When determining proper modal controllability norms of a system, both placement and the output power range of the actuator can effect the norm. When the concern for obtaining a norm is to determine placement of actuators of equal power capability, the best method for determining relative controllability norms comes from the simultaneous scaling of B_f and $\mathbf{u}_z$ such that the columns of B_f are normalized with magnitude 1.[9] i.e.

$$B_f \mathbf{u}_z = B_f P_n^{-1} P_n \mathbf{u}_z = B_{fn} \mathbf{u}_n \tag{39}$$

where P_n is a diagonal matrix containing the norms of the columns of B_f. When B_{fn} and u_n are used instead of B_f and u, the result is that B_{mp} is properly normalized to be B_{mpn}. The controllability norm of the qth mode from the rth input is then

$$\mathbb{C}_{q,r} |B_{mpn_{q,r}}| \tag{40}$$

where $B_{mpn_{q,r}}$ is the element of B_{mpn} in the qth row and the rth column. It is also possible to define the controllability norm of the qth mode to a set R of inputs where R is a list of numbers corresponding to those inputs. This is given by

$$\mathbb{C}_{q,R} = \det(B_{mpn_{q,R}} B_{mpn_{q,R}}^T)^{(1/2N_q)} \tag{41}$$

where $B_{mpn_{q,r}}$ is a matrix comprised of the set R rows of B_{mpn_q}.[10] Note that when the qth natural frequency is a repeated frequency, B_{mpn_q} represents the normalized B matrix from the qth equation of Equations (35). When the controllability norm of a mode from a certain actuator or set of actuators is zero, the mode is uncontrollable from that set of actuators.

When considering placement of actuators of varied power capabilities, a different weighting of $B_{mpn_{q,R}}$ is advisable. A reasonable weighting scheme is to weight B_{mp} and u_z as in Equation (39) but choose P_n such that

$$P_{n_{i,j}} = \frac{1}{\max(u_i)} \tag{42}$$

where $\max(u_i)$ represents the maximum output of the ith actuator.

By using Equation (42) the controllability norms are based on how much force each actuator can apply to each mode as well as where the actuator is placed. Thus a poorly placed more powerful actuator can have the same control impact on a given mode as a well placed weaker actuator. This allows sizing of actuators via modal controllability norms that was not previously possible.

Note that these norms, which are the same as the norms defined by Hamdan and Nayfeh[9] and Hughes and Skelton[10], are derived from the decoupling of the system into modal coordinates (as by Hughes and Skelton) but using normalized rows and columns (as by Hamdan and Nayfeh). Hamdan and Nayfeh[9] state that these results are valid for systems with complex modes, but ignore the decoupling difficulty demonstrated in Equation (24). This equation clearly shows that there is more going on in the controllability norms than the immediately preceding section can define. It is clear from (24) that the concept of complex modes is violated when a force is applied because the top half of the state equation, which represents an identity statement, is violated. Thus $w_2 \neq \dot{w}_1$ where w_1 and w_2 are the first and second halves of the modal state vector respectively. For systems with slightly complex modes, this may not present a practical problem. But for systems with "very" complex modes, defining modal norms of any sort for a forced system is risky at best.

3.5.4. Modal Observability of a Linear System

The ability to sense each mode of interest is also of concern when attempting vibration control. Optimal sizing and placement of sensors improves the signal to noise ratio and thus enables better control.

For a system written in the form (15), the necessary and sufficient conditions for observability of the system is

$$\text{rank}[C\ CA\ CA^2\ \cdots\ CA^{2N-1}]^T = 2N \tag{43}$$

For vibration suppression, a better way to consider observability is to determine the observability of a mode and identify a measure or norm quantifying how observable each mode is. Consider a system of the form (15) with modal damping. If we make the modal substitution $\mathbf{z} = P_{fp}\mathbf{w}$ into the observation equation, we get

$$\mathbf{y} = C_p P_{fp}\mathbf{w} = C_{pm}\mathbf{w} \tag{44}$$

Observability can be affected by both sensor placement and the sensitivity of the sensor. For sensors of equal sensitivity, the best method for determining norms is to simultaneously scale $\mathbf{y}$ and C_{pm} such that the rows of C_{pm} are normalized with magnitude 1^9, i.e.,

$$P_{cn}^1\mathbf{y} = P_{cn}^{-1}C_{pm}\mathbf{w}$$
$$\mathbf{y}_n = C_{pmn}\mathbf{w} \tag{45}$$

where P_{cn} is a diagonal matrix consisting of the norms of the rows of the matrix C_p. When the normalized variables are used, the elements of C_{pmn} represent the observability norms of the system.[9,10] The observability norm of the qth modal degree of freedom from the rth sensor is then

$$\varnothing_{q,r} = |C_{pmn_{r,q}}| \tag{46}$$

where $C_{pmn_{r,q}}$ is the element of C_{pmn} in the rth row and the qth column. It is also possible to define a gross observability norm of the qth degree of freedom to a set R sensors. This is given by

$$\varnothing_{q,R} = (C_{pmn_{R,q}}^T C_{pmn_{R,q}})^{1/2} \tag{47}$$

where $C_{pmn_{R,q}}$ is a row vector comprised of the Rth rows of the qth column of C_{pmn}.

When considering placement of sensors of varied sensitivities, a different scheme should be used. (What is meant here by the sensor is all mechanics and electronics between the physical responses being sensed and the voltage read by the data acquisition system.) For instance, although a very sensitive sensor may be poorly placed for a given mode, it may work just as well for observing the mode as a well placed but less sensitive sensor. Assuming noise is caused by A/D conversion, cable noise, or amplifier noise, either scenario could be just as good. What is important in choosing a sensor is that the range of the response being sensed is within an order of magnitude of the range of the sensor. If the sensor is not sensitive enough, the A/D conversion will not be smooth and electrical noise can dominate the signal. A more sensitive sensor will yield a smoother A/D conversion and less electrical noise. Of course the range of the responses being measured must be within the range of the sensor's capabilities. For this purpose, it is proposed that sensitivity be incorporated into the sensitivity norm by normalizing $\mathbf{y}$ and C_{pm} such that the elements of $\mathbf{y}$ have a maximum value of 1, i.e.

$$P_{cn}^{-1}\mathbf{y} = P_{cn}^{-1}C_{pm}\mathbf{w}$$
$$\mathbf{y}_n = C_{pmn}\mathbf{w}$$

(48)

where P_{cn} is a diagonal matrix containing the sensitivities of the sensors such that $P_{i,i}$ is the sensitivity of the ith sensor. Note that unlike the actuators, we do not use the maximum range. Often the maximum range is a physical or structural limitation and may not necessarily correspond to the sensitivity. For sensors of similar design with variations only in electronics, the maximum range and sensitivity usually do have a one to one relationship.

3.5.5. Modal Control of a Linear System

For modal control, we choose output feedback of the form $\mathbf{u}_z = G\mathbf{y}$ where G is a $N \times 2N$ matrix which can be partitioned into two $N \times N$ matrices G_p and G_v such that $G = [G_p \ G_v]$. Equation (20) then becomes

$$\dot{\mathbf{w}} = \begin{bmatrix} 0 & I \\ -\Omega^2 & -2Z\Omega \end{bmatrix}\mathbf{w} + \begin{bmatrix} 0 \\ P_s^{-1}B_f G(C_p P_{fp}\mathbf{w} + C_v P_{fp}\dot{\mathbf{w}}) \end{bmatrix}$$

(49)

As can be seen in Equation (49), the simplest solution arises if $\dot{\mathbf{w}}$ does not influence the output $\mathbf{y}$. Thus C_v is assumed to be equal to zero. Also, C_p is assumed to be of the form

$$C_p = \begin{bmatrix} C_{p_{1,1}} & 0 \\ 0 & C_{p_{2,1}} \end{bmatrix}$$

(50)

These assumptions are equivalent to the assumptions made in the second order modal control derivation of Inman[11], Hughes and Skelton[10] and Hamdan and Nayfeh.[9] Equation (49) can then be written as

$$
\dot{\mathbf{w}} = \begin{bmatrix} 0 & I \\ -\Omega^2 & -2Z\Omega \end{bmatrix} \mathbf{w} + \begin{bmatrix} 0 & 0 \\ -P_s^{-1}B_fG_pC_{p_{1,1}}P_s & -P_s^{-1}B_fG_vC_{p_{2,2}}P_s \end{bmatrix} \mathbf{w}
$$
(51)

It is clear from Equation (51) that for the modal control to be able to place the natural frequencies and damping ratios properly, the matrices $C_{p_{1,1}}$ and $C_{p_{2,2}}$ must be designed properly. As stated in Bellman[3], if two real symmetric matrices commute then they will be simultaneously diagonalized by a real orthogonal transformation. Thus, if $B_fG_pC_{p_{1,1}}$ and $B_fG_vC_{p_{2,2}}$ are symmetric,

$$
B_fG_pC_{p_{1,1}}M^{-1}K = KM^{-1}B_fG_pC_{p_{1,1}}
$$
(52)

and

$$
B_fG_vC_{p_{2,2}}M^{-1}K = KM^{-1}B_fG_vC_{p_{2,2}}
$$
(53)

then the matrices $P_s^{-1}B_fG_pC_{p_{1,1}}P_s$ and $P_s^{-1}B_fG_vC_{p_{2,2}}P_s$ are diagonal.[1,3,5]

If these conditions are satisfied, then Equation (51) simplifies to N single degree of freedom first order systems with the form

$$
\begin{bmatrix} \dot{w}_i \\ \ddot{w}_i \end{bmatrix} = \begin{bmatrix} 0 & 1 \\ -\omega_i^2 - \alpha_i & -2\zeta_i\omega_i - \beta_i \end{bmatrix} \begin{bmatrix} w_i \\ w_i \end{bmatrix}
$$
(54)

The variables α_i and β_i can then be chosen to give the desired modal frequencies and damping ratios. While this appears to be quite simple, in reality the desired values α_i and β_i cannot usually be implemented independently. The design problem is then to choose B_f, G_p, G_v, $C_{p_{1,1}}$, and $C_{p_{2,2}}$ such that

$$
P_s^{-1}B_fGC_{p_{1,1}}P_s = \mathrm{diag}[\alpha_i]
$$
(55)

$$
P_s^{-1}B_fGC_{p_{2,2}}P_s = \mathrm{diag}[\beta_i]
$$
(56)

Since P_s is characteristic of the system, it is presumed to be known apriori such that Equations (55) and (56) can be written

$$
B_fGC_{p_{1,1}} = P_s \, \mathrm{diag}[\alpha_i]P_s^{-1}
$$
(57)

$$
B_fGC_{p_{2,2}} = P_s \, \mathrm{diag}[\beta_i]P_s^{-1}
$$
(58)

As can be seen in Equations (55) and (56), not only must the control law be chosen properly, but the placement of the actuators and sensors must be made carefully too so that Equations (55) and (56) may be satisfied.

3.6. NONLINEAR MODAL ANALYSIS, OBSERVABILITY, AND CONTROLLABILITY

The concept that nonlinear modes with nonlinear modal equations exist for some set of nonlinear systems has been accepted intuitively by many for quite some time. It wasn't until 1962 when Rosenberg[23] presented the first paper on nonlinear normal modes that it became possible to solve even the simplest nonsimilar normal mode system. Many perturbation methods can approximate the deviation of a nonlinear mode from a corresponding linear mode. However, only the Shaw and Pierre method[25,26] utilizes the definition of nonsimilar nonlinear normal modes as invariant manifolds. This method allows the straight forward computation of nonlinear normal modes and their corresponding mode shapes. Although other methods have been proposed[13,17,18,37], they have not been demonstrated to be applicable to nonconservative systems, specifically those where the damping is non-modal. This method also readily lends itself to programming using algebraic manipulation packages such as Mathematica® and MACSYMA®. The following is an overview of what is a nonlinear normal mode and how one can be determined via the method proposed by Shaw and Pierre.

3.6.1. An Overview of Nonlinear Normal Modes

Most dynamicists are, in general, comfortable with the notion of a linear normal mode. The dynamics of a system which is moving in a damped linear normal mode can be written in the following vector form.

$$\mathbf{x}(t) = \mathbf{x}|_{t_0} e^{-t/\tau} \sin(\omega t + \Phi) \qquad (59)$$

where $\mathbf{x}|_{t_0}$ is the mode shape or the spatial part of the modal dynamics, t is time, τ is the time constant of the mode (∞ for the undamped system), ω is the natural frequency of the mode, and Φ is a phase angle.

Unlike the linear normal mode case, nonlinear normal modes are not quite as widely known. Nonlinear normal modes are defined to be either "similar" or "nonsimilar". A similar nonlinear normal mode is one in which the mode shape is not dependent on the modal amplitude, and thus is "similar" to a linear mode. The following example illustrates this.

Consider the symmetric spring mass system of Figure 1. All masses have a mass of 1 and the springs have cubic stiffness such that $f = x^3$. The equations of motion for the system are given by

$$\ddot{x}_1 = x_2^3 - 2x_1^3 \qquad (60)$$

J.C. SLATER AND G. AGNES

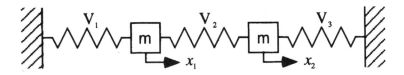

Figure 1. A 2-DOF Linear System.

and

$$\ddot{x}_2 = x_1^3 - 2x_2^3 \tag{61}$$

Here it is assumed that the system has similar normal modes such that $x_2 = ax_1$. Substituting this assumption into the equations of motion yields

$$\ddot{x}_1 = (a^3 - 2)x_1^3 \tag{62}$$

and

$$\ddot{x}_1 = \frac{1}{a}(1 - 2a^3)x_1^3 \tag{63}$$

Setting the right hand sides of these expressions equal to each other yields

$$a^4 + 2a^3 - 2a - 1 = 0 \tag{64}$$

The roots are then $a = -1$ (3 times) and $a = 1$. The mode shapes are then given by

$$\begin{bmatrix} x_1 \\ x_1 \end{bmatrix} \quad \text{and} \quad \begin{bmatrix} x_1 \\ -x_1 \end{bmatrix} \tag{65}$$

As mentioned earlier, the relationship between the modal coordinate and the mode can be chosen rather arbitrarily as long as the choice allows the determination of the modal equations of motion. A different relationship will cause a different modal equation to be found, but the dynamics of each degree of freedom when moving in the mode will remain the same. Selecting the relationship between the modal coordinate u and the mode shapes to be $X_1 = u$ and substituting both mode shapes into the first equation of motion yields the two modal equations

$$\ddot{u}_1 = -u_1^3 \tag{66}$$

and

$$\ddot{u}_2 = -3u_2^3 \tag{67}$$

where u_1 and u_2 are the first and second modal coordinates, respectively.

The existing definition[24,33] for a nonlinear normal mode states that a discrete N degree of freedom system is oscillating in a normal mode if all of the motions are periodic of the same period, all of the coordinates reach their extreme values at the same time, and at any instant in time the coordinate displacements can be related by a functional relationship of the form

$$x_i = f_i(u(t)) \tag{68}$$

where x_i is the ith displacement, u is the modal displacement, and f_i is the relationship relating them, then the system is moving in a nonlinear mode. This definition is restricted to undamped systems. Motions of a system freely oscillating in a modally damped mode will not, in general, be periodic and of the same period. A simple example of a system oscillating in a modally damped mode in a single degree of freedom mass spring system with velocity squared damping. Each cycle of the motion will exhibit a slightly different period during the decay. Also, linear systems which exhibit complex modes are not normal modes by this definition since the motion of the system is not in unison.

Likewise, Shaw and Pierre[25] define a nonlinear mode as having the following properties: "it passes through a stable equilibrium point of the system and, at that point, it is tangent to a plane which is an eigenspace of the system linearized about that equilibrium." This is vastly improved over previous definitions, since it applies to damped systems as well.

In linear systems, modes are categorized as real or complex. With the advent of nonlinear modal analysis, it is advantageous to likewise categorize nonlinear modes. The following set of definitions has been proposed by Slater.[29]

Definition one: A system is oscillating in a normal mode if the motion of any point $(u_{\hat{x}}, u_{\hat{y}}, u_{\hat{z}})$ in three dimensional space $(\hat{x}, \hat{y}, \hat{z})$ can be described by the equation

$$\begin{bmatrix} u_{\hat{x}}(\hat{x}, \hat{y}, \hat{z}) \\ u_{\hat{y}}(\hat{x}, \hat{y}, \hat{z}) \\ u_{\hat{z}}(\hat{x}, \hat{y}, \hat{z}) \end{bmatrix} = \begin{bmatrix} f_{\hat{x}}(u(t), v(t)) \\ f_{\hat{y}}(u(t), v(t)) \\ f_{\hat{z}}(u(t), v(t)) \end{bmatrix} \tag{69}$$

where $(\hat{x}, \hat{y}, \hat{z})$ represents the location of the point on the structure in three dimensional space, $u_{\hat{x}}$, $u_{\hat{y}}$ and $u_{\hat{z}}$ represent the deflections in the x, y, and z directions, $f_{\hat{x}}$, $f_{\hat{y}}$ and $_{\hat{z}}$ relate the deflections to the modal coordinates u and v. This represents the two-dimensional invariant manifold described by Shaw and Pierre.[25]

Definition two: If the function f_i relating the displacements $u_{\hat{i}}$ to the modal coordinates are linear and the modal equations in u and v are linear then the mode is a similar linear normal mode, often referred to as a linear mode.

Definition three: If the functions f_i relating the displacements u_i to the modal coordinates are linear and the modal equations in u and v are nonlinear then the mode is a similar nonlinear normal mode. This corresponds to the definition put forth by Rosenberg[24] that if the modal curves corresponding to a nonlinear normal mode are straight, then the nonlinear mode is called "similar."

Definition four: If the functions f_i relating the displacements u_i to the modal coordinates are nonlinear then the mode is a nonsimilar nonlinear normal mode. This corresponds to the definition put forth by Rosenberg[24] that if the modal curves corresponding to a nonlinear normal mode are not straight, then the mode is called a "nonsimilar" nonlinear normal mode.

Definition five: If the trajectory of a nonlinear normal mode passes through static equilibrium it is an equal phase nonlinear normal mode. For a linear normal mode this type of mode is usually called a "real" mode.

Definition six: If the trajectory of a nonlinear normal mode does not pass through static equilibrium it is a non-equal phase nonlinear normal mode.

In addition to the definitions proposed by Slater[29], the following is added:

Definition seven: If the functions F_i relating the displacements u_i to the modal coordinates are linear, depend also on the velocities, and the modal equations in u and v are nonlinear, then the mode is a non-equal phase similar nonlinear normal mode.

The following sections describe the method developed by Shaw and Pierre for determining nonlinear normal modes. As will become evident, the method allows for the solution of nonlinear normal modes which were previously undefined by the Rosenberg definition.

3.6.2. Normal Modes of Nonlinear Systems

The equations of motion for a nonlinear oscillatory system may be written in state space form as

$$\dot{x}_i = y_i \tag{70}$$

$$\dot{y}_i = \tilde{A}_i(\mathbf{x}, \mathbf{y}) \tag{71}$$

where x_i is the displacement of the ith degree of freedom, y_i is the velocity of the ith point, and $\tilde{A}_i(\mathbf{x}, \mathbf{y})$ is the function describing the dynamics of the ith degree of freedom. The variables $\mathbf{x}$ and $\mathbf{y}$ are the vectors of displacements and velocities. The assumption is made that the motion can take place in a two dimensional invariant manifold. That is, the motion of each degree of freedom may be written as a function of a modal displacement, u, and a modal velocity, v. The motion of the system along a single invariant manifold is then described by the equations

$$x_i = X_i(u, v) \tag{72}$$

and

$$\tag{73}$$

$$y_i = Y_i(u, v) \tag{74}$$

where

$$\tag{75}$$

$$i = 1, 2, 3, \cdots, N \tag{76}$$

and the functions X_i and Y_i relate the physical displacements (x_i) and velocities (y_i) to the modal displacement and velocity. Taking the derivative of Equations (72) and (74) using the chain rule yields

$$\dot{x}_i = \frac{\partial X_i}{\partial u}\dot{u} + \frac{\partial X_i}{\partial v}\dot{v} \tag{77}$$

and

$$\dot{y}_i = \frac{\partial Y_i}{\partial u}\dot{u} + \frac{\partial Y_i}{\partial v}\dot{v} \tag{78}$$

Setting the right hand sides of Equations (77) and (78) equal to the right hand sides of Equations (70) and (71) yields

$$y_i = \frac{\partial X_i}{\partial u}\dot{u} + \frac{\partial X_i}{\partial v}\dot{v} \tag{79}$$

and

$$\tilde{A}_i(\mathbf{x}, \mathbf{y}) = \frac{\partial Y_i}{\partial u}\dot{u} + \frac{\partial Y_i}{\partial v}\dot{v} \tag{80}$$

Substituting (72) and (74) into (79) and (80) gives the equations of motion entirely in terms of the modal coordinates u, and v as (81)

$$Y_i(u, v) = \frac{\partial X_i}{\partial u}\dot{u} + \frac{\partial X_i}{\partial v}\dot{v} \tag{81}$$

and

$$\tilde{A}_i(\mathbf{X}(u, v), \mathbf{Y}(u, v)) = \frac{\partial Y_i}{\partial u}\dot{u} + \frac{\partial Y_i}{\partial v}\dot{v} \tag{82}$$

where $\mathbf{X}(u, v) = \mathbf{x}$ and $\mathbf{Y}(u, v) = \mathbf{y}$.
    By definition

$$\dot{u} = v \tag{83}$$

Selecting $x_i = u$ gives

$$\dot{v} = \ddot{u} = \ddot{x}_1 = \dot{y}_1 = \tilde{A}_1(\mathbf{x}, \mathbf{y}) \tag{84}$$

Substituting (83) and (84) into (81) and (82) and using $\mathbf{X}(u, v) = \mathbf{x}$ and $\mathbf{Y}(u, v) = \mathbf{y}$ yields $2N$ partial differential equations for $\mathbf{X}(u, v)$ and $\mathbf{Y}(u, v)$

$$Y_i(u, v) = \frac{\partial X_i(u, v)}{\partial u} v + \frac{\partial X_i(u, v)}{\partial v} \tilde{A}_1(\mathbf{X}_i(u, v), \mathbf{Y}_i(u, v)) \tag{85}$$

and

$$\tilde{A}_i(\mathbf{X}(u, v), \mathbf{Y}_i(u, v)) = \frac{\partial Y_i(u, v)}{\partial u} v + \frac{\partial Y_i(u, v)}{\partial v} \tilde{A}_1(\mathbf{X}_i(u, v), \mathbf{Y}_i(u, v)) \tag{86}$$

Note that for $i = 1$, Equations (85) and (86) are satisfied identically since

$$X_1(u, v) = u, \tag{87}$$

and $\tag{88}$

$$Y_1(u, v) = v \tag{89}$$

In general, these equations are difficult to solve. However, they do lend themselves to a power series solution. Once they have been solved for the $X_i(u, v)$ and $Y_i(u, v)$, the dynamics of the invariant manifold (or modal dynamics) can be generated by substituting the $X_i(u, v)$ and $Y_i(u, v)$ in for x_i and y_i in the first pair of equations of motion (Substitution into the other equations of motion is correct, but more difficult). This results in the modal equations

$$\dot{u} = v \tag{90}$$

$$\dot{v} = \tilde{A}_{mi}(u, v) \tag{91}$$

where u and v are the modal coordinates and $X_i(u, v)$ and $Y_i(u, v)$ represent the mode shapes as functions of the modal displacement and velocity. In general, at each equilibrium point there are N solutions for the $X_i(u, v)$ and $Y_i(u, v)$ and N sets of equations of the form given in (90) and (91).

In some cases, such as systems which possess certain symmetries, Equations (87) and (89) can be solved exactly (usually for systems exhibiting similar normal modes). In this case, exact representations of the mode shapes and their associated modal equations can be obtained. This procedure allows the solution of systems with both similar and non-similar modes as well as systems exhibiting velocity dependence.

Consider the following system

$$M\ddot{x} + Kx = 0 \tag{92}$$

where

$$M = \begin{bmatrix} 9 & 0 \\ 0 & 1 \end{bmatrix}, \quad K = \begin{bmatrix} 27 & -3 \\ -3 & 3 \end{bmatrix} \tag{93}$$

The equation of motion in state space form is

$$\begin{bmatrix} \dot{x}_1 \\ \dot{x}_2 \\ \dot{y}_1 \\ \dot{y}_2 \end{bmatrix} = \begin{bmatrix} 0 & 0 & 1 & 0 \\ 0 & 0 & 0 & 1 \\ -3 & \dfrac{1}{3} & 0 & 0 \\ 3 & -3 & 0 & 0 \end{bmatrix} \begin{bmatrix} x_1 \\ x_2 \\ y_1 \\ y_2 \end{bmatrix} \tag{94}$$

Thus

$$A_1 = -3x_1 + \frac{1}{3}x_2, \quad A_2 = 3x_1 - 3x_2 \tag{95}$$

The first degree of freedom is arbitrarily chosen to be the modal coordinate. Thus $x_1 = u$ and $y_1 = v$. Assuming a solution of the form $x_2 = au$ and $y_2 = bv$, and applying Equation (85) for $i = 2$ yields

$$bv = av \tag{96}$$

Recall that Equations (85) and (86) are indentically satisfied for $i = 1$. Next, applying Equation (86) yields

$$3u - 3au = b\left(-3u + \frac{1}{3}au\right) \tag{97}$$

Solving Equations (96) and (97) yields $a = b$ and $b = \pm 3$. The mode shapes are then $x_2 = 3x_1$ and $x - 2 = -3x_1$.

The approximate "superposition" proposed by Shaw and Pierre is a nonlinear extension of the usual linear concept, and is not strictly the same as that in a linear system. It is, in essence, a nonlinear coordinate transformation which simplifies a large number of second order Equations (or sets of two first order linear equations) of motion into simpler independent second order equivalent nonlinear systems that can be recombined to yield the total system dynamics.

The geometry behind the idea is quite simple in concept. In linear systems, one has planar eigenspaces and the general response can be broken down into modal components via a linear projection of the response onto the eigenspaces and reassembled by a linear recombination of the modal responses. When nonlinearities are present, the modal subspaces are not linear, but curved, and the required coordinate transformations are usually nonlinear. Also, the equations of motion transformed into modal space are only independent in the sense that motion can occur such that the states of the system are restricted to a single manifold. The equations are coupled (slightly) when motion takes place on multiple manifolds.

As stated in the previously, in some cases an exact solution for the modal dynamics is possible. However, in general an exact solution is not possible. In these cases the approach taken is to assume a power series solution, substitute the assumed solution into Equations (85) and (86) and solve for the coefficients by matching the coefficients of like powers to solve for the coefficients of the power series.

This approach is based on a power series expansion about the equilibrium state. The development in the previous section makes no reference to an equilibrium point. Therefore, the first step of the analysis is to solve for the equilibrium state. This is done by solving for the states x_0 and $\mathbf{y} = \mathbf{0}$ such that

$$\tilde{A}_i(\mathbf{x}_0, \mathbf{0}) = 0 \quad 1 = 1, 2, 3, \ldots, N \tag{98}$$

Once this is done, we can apply the power series expansions about the equilibrium of interest. Next it is assumed that the nonlinear normal modes of the system can be represented by a power series in u and v. Thus, using the notation of Shaw and Pierre[25]

$$X_i(u, v) = a_{1i}u + a_{2i}v + a_{3i}u^2 + a_{4i}uv + a_{5i}v^2 + a_{6i}u^3 \cdots \tag{99}$$

$$Y_i(u, v) = b_{1i}u + b_{2i}v + b_{3i}u^2 + b_{4i}uv + b_{5i}v^2 + b_{6i}u^3 \cdots \tag{100}$$

In vector form this series can be represented as

$$\begin{bmatrix} x_1 \\ y_1 \\ x_2 \\ y_2 \\ \vdots \\ x_N \\ y_n \end{bmatrix} = \left\{ \begin{bmatrix} 1 & 0 \\ 0 & 1 \\ a_{12} & a_{22} \\ b_{12} & b_{22} \\ \vdots & \vdots \\ a_{1N} & a_{2N} \\ b_{1N} & b_{2N} \end{bmatrix} + \begin{bmatrix} 0 & 0 \\ 0 & 0 \\ a_{32}u + a_{42}v & a_{52}v \\ b_{32}u + b_{42}v & b_{52}v \\ \vdots & \vdots \\ a_{3N}u + a_{4N}v & a_{5N}v \\ b_{3N}u + b_{4N}v & b_{5N}v \end{bmatrix} + \right.$$

$$\left. \begin{bmatrix} 0 & 0 \\ 0 & 0 \\ a_{62}u^2 + a_{82}v^2 & a_{72}u^2 + a_{92}v^2 \\ b_{62}u^2 + b_{82}v^2 & b_{72}u^2 + b_{92}v^2 \\ \vdots & \vdots \\ a_{6N}u^2 + a_{8N}v^2 & a_{7N}u^2 + a_{9N}v^2 \\ b_{6N}u^2 + b_{8N}v^2 & b_{7N}u^2 + b_{9N}v^2 \end{bmatrix} + \cdots \right\} \begin{bmatrix} u \\ v \end{bmatrix} \tag{101}$$

or more compactly

$$\mathbf{z} = \mathbf{m} \begin{bmatrix} u \\ v \end{bmatrix} = [\mathbf{m}_0 + \mathbf{m}_1(u, v) + \mathbf{m}_2(u, v) + \cdots] \begin{bmatrix} u \\ v \end{bmatrix} \tag{102}$$

where

$$\mathbf{z} = [x_1, y_1, x_2, y_2, \cdots, x_N, y_N]^T \tag{103}$$

and $\mathbf{m}_0$, $\mathbf{m}_1$, and $\mathbf{m}_2$ are $2N \times 2$ matrices. The matrix $\mathbf{m}_0$ is the linear component of the nonlinear modes and $\mathbf{m}_2$ and $\mathbf{m}_3$ represent the quadratic and cubic terms respectively. This representation may be cumbersome at this point, but it is helpful in understanding upcoming coordinate transformations. The matrix $\mathbf{m}_0$ represents the mode shapes common to linear systems. For a linear system, all other matrices $\mathbf{m}_j$ would be zero. For the linear undamped or modally damped system, the cross terms a_{2i} and b_{1i} are zero while $a_{1i} = b_{2i}$ is the usual amplitude ratio relating the ith degree of freedom to the modal amplitude. For a normally damped system, the cross terms are generally non-zero and thus represent the effect of complex modes in terms of real numbers.

The nonlinear terms represent the bending of the modal subspace away from the linear solution. Their associated coefficients can all be zero in which case the system is either linear or the mode is a similar normal mode. These terms represent the amplitude and velocity dependence of the restoring forces of the nonlinear system.

The series representations of the nonlinear normal modes is now substituted into Equations (85) and (86). The result is a set of $2N - 2$ equations which contain the coefficients of the Taylor series of the representation of the forces of Equation (71) as known quantities and the coefficients of the nonlinear normal modes as unknowns. The coefficients of the terms in Equations (99) and (100) are found by gathering terms of like powers in u and v and equating their coefficients. These equations can be solved sequentially, starting with the linear equations, to desired order. The equations for the linear coefficients are quadratic in the unknowns a_{2j}, a_{2j}, b_{2j}, and b_{2j} and generally have N real solutions, one for each mode. The equations for the quadratic coefficients depend on the linear coefficients and examination of the equation shows that they are *linear* in the unknown quadratic coefficients, and thus generally have a unique solution for each mode. The equations for the cubic coefficients depend on the linear and quadratic coefficients and examination of these equations yields that these are linear in the cubic coefficients. Thus each of the higher order coefficients may be found from knowledge of the lower order coefficients and the solution of a set of linear equations. As Shaw and Pierre point out, this methodology will be easy to incorporate into computational schemes, even for large scale systems. Boivin, Pierre, and Shaw[1] have shown that solving the nonlinear equations for the linear terms is unnecessary, since they simply represent the linear mode shapes. Applying a linear eigenvalue solution method to the linear equations, and projecting the nonlinear equations onto the linear modes yields a much more robust solution.

This technique is not applied in this work since they are unnecessary for the examples given. It could, however, be easily incorporated into the following theory.

The solutions of these equations result in series approximations for the N nonlinear normal modes. The approximation of the dynamic equations for each mode can be found by using the Taylor series approximation of $\tilde{A}_i(\mathbf{x}, \mathbf{y})$ in Equation (71) and substituting the power series approximation of $\mathbf{X}(u, v)$ and $\mathbf{Y}(u, v)$ into Equations (70) and (71) for $\mathbf{x}$ and $\mathbf{y}$. Substitution of the N different nonlinear mode solutions into Equations (70) and (71) yields N single degree of freedom nonlinear modal equations. Each equation represents the equation of motion of the system on the corresponding invariant manifold.

Now the complete nonlinear modal matrix M can be assembled from the modal vectors $\mathbf{m}$. The modal matrix $M(\mathbf{w})$ is then

$$M(\mathbf{w}) = [{}_1\mathbf{m}\ {}_2\mathbf{m}\ {}_3\mathbf{m} \cdots {}_N\mathbf{m}] \tag{104}$$

where ${}_i\mathbf{m}$ represents the modal vector for the ith mode and $\mathbf{w} = [u_1, v_1, u_2, v_2, = cdotsu_N, v_N]^T$ where u_i and v_i are the modal displacement and velocity respectively for the ith mode. The complete transformation from modal to physical coordinates can now be written

$$\mathbf{z} = M(\mathbf{w})\mathbf{w} \tag{105}$$

Next, the matrix $M(\mathbf{w})$ can be subdivided in the same manner as the vector $\mathbf{m}(\mathbf{w})$ was in Equation (102). This gives

$$M(\mathbf{w}) = M_0 + M_1(\mathbf{w}) + M_2(\mathbf{w}) \tag{106}$$

If one were to write out $M(\mathbf{w})$ in general form, the coefficient a_{ij} and b_{ij} will in general be different for each mode. Shaw and Pierre suggest that the notation a_{ijk} and b_{ijk} be used where the index k represents the mode number. The importance of this form for $M(\mathbf{w})$ will become apparent in the next section.

3.6.3. Transformation from Physical to Nonlinear Modal Coordinates

It must be made clear that these transformations are an extension of linear theory but superposition does not apply in the linear sense. It is similar to the well known linear transformations only in that it allows complex system to be viewed as a number of simpler systems. These systems can be solved and their solutions combined to yield the total system dynamics.

For linear systems, the transformation between modal and physical coordinates is given by the matrix M_0. The transformation from the physical to the modal coordinates for the nonlinear system is not the same as the transformation from the modal to the physical coordinates due to the modal amplitude dependence of the transformation. For the transformation from modal to physical coordinates, it is assumed that the modal amplitudes are known. Since the transformation matrix is written in terms of the modal amplitudes, the transformation can be carried out (see Equation (105)). In making the transformation from physical to modal coordinates we cannot simply invert the matrix $M(\mathbf{w})$ in order to accomplish the transformation because we do not know the modal amplitudes and therefore cannot evaluate $M(\mathbf{w})$. The transformation method developed by Shaw and Pierre[25] is shown below.

Index notation and the implicit summation notation are employed here for deriving the inverse of the modal transformation.

The forward coordinate transformation is written

$$z_k = M_{1kj} w_j M_{2kjr} w_j w_r + M_{3kjrs} w_j w_r w_s + \ldots \tag{107}$$

and its inverse is expressed as

$$w_m = Q_{1mn} z_n + Q_{2mno} z_n z_o + Q_{3mnop} z_n z_o z_p + \ldots \tag{108}$$

While the forward coordinate transformation is obtained from application of the invariant manifolds, the inverse must be derived from the forward transformations. Substitution of Equation (107) into Equation (108) and expanding to third order terms yields

$$w_m = Q_{1mn} M_{1nj} w_j + Q_{1mn} M_{2njr} w_j w_r + Q_{2mno} M_{1nj} M_{1oi} w_j w_i$$
$$+ Q_{1mn} M_{3njrs} w_j w_r w_s + Q_{2mno} M_{1nj} M_{2oab} w_j w_a w_b \tag{109}$$
$$Q_{2mno} M_{1oi} M)_{njr} w_i w_j w_r + Q_{3mnop} M_{1oi} M_{1pa} w_j w_i w_a + \ldots$$

Gathering like powers of w, the linear terms yield

$$Q_{1mn} M_{1nj} = \delta_{mj} \tag{110}$$

This is the expected result the forward and backward linear coordinate transformations are related by a matrix inverse. The quadratic terms yield

$$Q_{1mn} M_{2njr} = -Q_{2mno} M_{1nj} M_{1oi} \tag{111}$$

which can be solved for Q_2 by utilizing the transpose of (110) and by post multiplying twice by Q_1 to eliminate the M_1 terms on the right hand side of (111). The result is

$$Q_{2miq} = -Q_{1mn} M_{2njr} Q_{1rq} Q_{1ji} \tag{112}$$

Likewise, the cubic terms must satisfy

$$Q_{3mnoq} M_{1nj} M_{1or} = -[Q_{1mn} M_{3njrm} + Q_{2mno} M_{1nj} M_{2orm} \\ + Q_{2mno} M_{1oj} M_{2nrm}] \tag{113}$$

which can be solved by three post multiplications of Q_1 to yield

$$Q_{3miqr} = [Q_{1mn} M_{3njrm} + Q_{2mno} M_{1nj} M_{2orm} \\ + Q_{2mno} M_{1oj} M_{2nrm}][Q_{1mnr} Q_{1rq} Q_{1ji}] \tag{114}$$

Note that this entire process required the a single matrix inverse, and extensive use of the result. The entire subsequent set of equations can be solved sequentially using simple tensor multiplications.

This results in the transformation

$$\mathbf{w} = \tilde{\mathbf{N}}(\mathbf{z}). \tag{115}$$

3.6.4. Transforming the Unforced Equations of Motion from Physical to Modal Coordinates

The equations of motion for a nonlinear system can be written

$$\mathbf{z} = \tilde{\mathbf{A}}(\mathbf{z}) = A(\mathbf{z})\mathbf{z} \tag{116}$$

For notational convenience, the transformation from physical to modal and modal to physical coordinates in Equations (105) and (115) can be written as

$$\mathbf{z} = \tilde{\mathbf{M}}(\mathbf{w}) \tag{117}$$

and

$$\mathbf{w} = \tilde{\mathbf{N}}(\mathbf{z}) \tag{118}$$

Using the chain rule, the time derivative of Equation (117) is

$$\dot{\mathbf{z}} = \frac{\partial \tilde{\mathbf{M}}(\mathbf{w})}{\partial \mathbf{w}} \dot{\mathbf{w}} \tag{119}$$

where $\partial \tilde{\mathbf{M}}(\mathbf{w})/\partial \mathbf{w}$ is the Jacobian of the vector $\tilde{\mathbf{M}}(\mathbf{w})$ with respect to $\mathbf{w}$. Substituting Equations (117) and (119) into the equations of motion yields.

$$\frac{\partial \tilde{\mathbf{M}}(\mathbf{w})}{\partial \mathbf{w}} \dot{\mathbf{w}} = A(\tilde{\mathbf{M}}(\mathbf{w}))\tilde{\mathbf{M}}(\mathbf{w}) \tag{120}$$

which can be premultiplied by the inverse of the Jacobian of the transformation to yield the equations of motion in modal coordinates

$$\dot{\mathbf{w}} = \left(\frac{\partial \tilde{\mathbf{M}}(\mathbf{w})}{\partial \mathbf{w}}\right)^{-1} A(\tilde{\mathbf{M}}(\mathbf{w}))\tilde{\mathbf{M}}(\mathbf{w}) \tag{121}$$

Using the series approximation methods described earlier, these equations can be constructed to desired order. Shaw and Pierre have noted that additional modal coupling terms can appear in these equation but that in simulations of example problems their effect is negligible when compared to other terms.

3.7. COORDINATE TRANSFORMATION RELATIONS, MODAL CONTROLLABILITY AND MODAL OBSERVABILITY

In this section, the concepts of modal coordinate transformation used for linear systems and developed by Shaw and Pierre[25] for unforced and unobserved systems is expanded to nonlinear modal systems with the inclusion of external forces and the observation equation. These transformation are used to derive nonlinear modal controllability norms and nonlinear modal observability norms which reduce to the norms derived in Section 3.5 for the linear case.

3.7.1. Coordinate Transformations

Unlike linear systems, nonlinear modal coordinate transformations are valid only when the motion of the system remains on a single manifold (or in a single mode). Off of the manifolds, the coordinate transformations are only approximate. Section 3.7.1.1 describes what effect these approximations have on using the nonlinear normal modal representation when the states of the system are off of a manifold. Section 3.7.1.2 describes the transformation of the nonlinear state equations from physical space to the nonlinear modal space, while Section 3.7.1.3 describes the inverse transformation (From nonlinear modal coordinates to physical coordinates).

3.7.1.1. Coordinate transformation relations

As in linear modal control, the first step in applying nonlinear modal control is the decoupling of the nonlinear system of equations into N single degree of freedom oscillators. Section 3.6.4 discusses transforming the unforced state equations of motion into modal coordinates by projecting the equations of

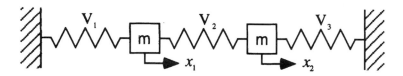

Figure 2. A 2-DOF Linear System.

motion onto the nonlinear modes. As Shaw and Pierre show, when using the coordinate transformation, the modal equations become coupled. This is due to the effect that unlike linear systems, the nonlinear mode shapes do not apply to the system when the states of the system are off of the invariant manifolds representing the mode shapes. In general, the more nonlinear the system is, the more coupling will exist between the modes.

Take for example the system of Figure 2 where m represents mass, V_i represents the potential energy of the ith spring, and x_1 and x_2 are the deflections. For a linear system, projecting the equations of motion onto the mode shapes completely decouples the system into two second order independent equations. Take $m_1 = m_2 = 1$ and

$$V_i = \frac{1}{2}(x_i - x_{i-1})^2 \tag{122}$$

where $x_0 = x_4 = 0$. The decoupling can be visualized by looking at Figure 3. The solid lines represent the two modes. The dashed line represents the mode shape of mode 1 when the amplitude of mode two is non-zero. The dot-dashed line represents the mode shape of mode 2 when the amplitude of mode 1 is non-zero. Assume now that the proper force can be applied to the system to maintain a constant non-zero amplitude for mode 1 and mode 2 is still free to move along the corresponding modal line. Because the system is completely decoupled, the dynamics of the second mode are unchanged.

Now consider the system of Figure 2 where all of the springs have equal but cubic stiffness, i.e.

$$V_i = \frac{1}{4}(x_i - x_{i-1})^4 \tag{123}$$

This system will exhibit similar nonlinear normal modes, as discussed in Section 6, identical to the linear system. Projecting the equations of motion onto the nonlinear normal modes yields two coupled second order nonlinear differential equations. Considering again that the proper forces can be applied to the system to maintain a constant non-zero amplitude for mode 1. Mode 2 is still free to move along the corresponding modal line. However, because of the coupling, the modal equation for mode 2 evaluated at the selected amplitude

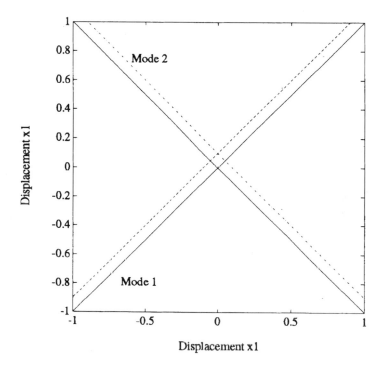

Figure 3. Similar normal modes.

for mode 1 is different than when it is evaluated at other amplitudes of mode 1. Thus the modal dynamics are different than at most other amplitudes of mode 1.

While no rigorous stability analysis is presented, a heuristic argument for the stability of the system can be made. Consider the ith nonlinear normal mode of an N degree of freedom system. The $N - 1$ modal equations other than the ith equation govern the motion of the system relative to the ith manifold. If those $N - 1$ equations are stable, then the ith nonlinear normal mode is attractive(i.e. all motion will converge to the ith manifold). Once the states of the system reach the ith manifold, then the stability of the system is determined by the stability of the ith mode. Thus the stability can be determined by evaluating the stability of the $N - 1$ modal equations at the initial condition and the stability of the ith manifold. Of course the constraint still remains that the initial condition does not cause the trajectory of the system to be pushed out of the "quasi-linear" space into a region in which bifurcation of modes or chaotic motion takes place.

The following three degree of freedom example demonstrates this theory. The partially decoupled equations of motion are given by

$$\dot{u}_1 = \tilde{A}_{m1}(\mathbf{u}, \mathbf{v}), \qquad \dot{v}_1 = \tilde{A}_{m2}(\mathbf{u}, \mathbf{v}) \tag{124}$$

$$\dot{u}_2 = \tilde{A}_{m3}(\mathbf{u}, \mathbf{v}), \qquad \dot{v}_2 = \tilde{A}_{m4}(\mathbf{u}, \mathbf{v}) \tag{125}$$

$$\dot{u}_3 = \tilde{A}_{m5}(\mathbf{u}, \mathbf{v}), \qquad \dot{v}_3 = \tilde{A}_{m6}(\mathbf{u}, \mathbf{v}) \tag{126}$$

where $\mathbf{u}$ and $\mathbf{v}$ are the modal coordinate vectors and $\tilde{A}_{mi}$ are the modal functions. The initial conditions are given as $u_i(0)$ and $v_i(0)$. Evaluating first two modal equations at the initial conditions gives

$$\dot{u}_i(0) = \tilde{A}_{m1}(\mathbf{u}, \mathbf{v})|_{u_2(0), v_2(0), u_3(0), v_3(0)} \tag{127}$$

$$\dot{v}_i(0) = \tilde{A}_{m2}(\mathbf{u}, \mathbf{v})|_{u_2(0), v_2(0), u_3(0), v_3(0)} \tag{128}$$

$$\dot{u}_2(0) = \tilde{A}_{m3}(\mathbf{u}, \mathbf{v})|_{u_1(0), v_1(0), u_3(0), v_3(0)} \tag{129}$$

and

$$\dot{u}_2(0) = \tilde{A}_{m4}(\mathbf{u}, \mathbf{v})|_{u_1(0), v_1(0), u_3(0), v_3(0)} \tag{130}$$

If these equations are stable and the ith nonlinear normal mode is stable, then the system should be stable from the initial conditions. By performing the preceding analysis on all of the modal equation, it may be possible to determine a region in state space within which the system is stable. However, this has not yet been proven.

3.7.1.2. *Transformation of nonlinear equations from physical to modal coordinates*

In Section 3.6.3, the unforced equations of motion were transformed into modal coordinates as described in Shaw and Pierre. Here we are interested in the approximate transformation of the state equation *and* the output equation into modal coordinates.

The nonlinear state equations can be written in physical coordinates as

$$\dot{\mathbf{z}} = \tilde{A}(\mathbf{z}) + B(\mathbf{z})\hat{\mathbf{u}}_z = A(\mathbf{z})\mathbf{z} + B(\mathbf{z})\hat{\mathbf{u}}_z \tag{131}$$

$$\hat{\mathbf{y}} = \tilde{C}(\mathbf{z}, \hat{\mathbf{u}}_z) \tag{132}$$

Note that the matrix $B(\mathbf{z})$ can be divided into two parts $B_{\text{odd}}(\mathbf{z})$ and $B_{\text{even}}(\mathbf{z})$. The variable $B_{\text{odd}}(\mathbf{z})$ represents the odd rows of the matrix $B(\mathbf{z})$ and the matrix $B_{\text{even}}(\mathbf{z})$ represents the even rows of $B(\mathbf{z})$. Since the odd rows of the state equation are the identity statements $\dot{u}_i = v_1$, the matrix $B_{\text{odd}}(\mathbf{z})$ is a zero matrix. Taking the derivative of Equation (105) with respect to time and applying the chain rule yields

$$\dot{\mathbf{z}} = \frac{\partial \dot{M}(\mathbf{w})}{\partial \mathbf{w}} \dot{\mathbf{w}} \tag{133}$$

Substituting (105) and (133) into (131) and (132) gives

$$\dot{\mathbf{w}} = \left(\frac{\partial \tilde{\mathbf{M}}(\mathbf{w})}{\partial \mathbf{w}}\right)^{-1} A(\tilde{\mathbf{M}}(\mathbf{w}))\tilde{\mathbf{M}}(\mathbf{w}) + \left(\frac{\partial \tilde{\mathbf{M}}(\mathbf{w})}{\partial \mathbf{w}}\right)^{-1} B(\tilde{\mathbf{M}}(\mathbf{w}))\hat{\mathbf{u}}_z \tag{134}$$
$$= \tilde{\mathbf{A}}_m(\mathbf{w}) + \hat{\mathbf{u}}_m(\mathbf{w})$$

and

$$\hat{\mathbf{y}} = \tilde{\mathbf{C}}(\tilde{\mathbf{M}}(\mathbf{w}), \hat{\mathbf{u}}_z) = \tilde{\mathbf{C}}_m(\mathbf{w}, \hat{\mathbf{u}}_z) \tag{135}$$

where $\hat{\mathbf{u}}_m(\mathbf{w})$ represents the amplitude dependent modal forces, $\tilde{\mathbf{A}}_m(\mathbf{w})$ represents the modal functions, and $\tilde{C}_m(\mathbf{w}, \hat{u}_g)$ represents the observation function in modal coordinates.

3.7.1.3. Transformation of nonlinear state equations from modal to physical coordinates

In transforming the equations from modal coordinates to physical coordinates, the same procedure is used as in the previous section. The equations of motion in modal coordinates are given by

$$\dot{\mathbf{w}} = \tilde{\mathbf{A}}_m(\mathbf{w}) + \hat{\mathbf{u}}_m(\mathbf{w}) \tag{136}$$

$$\hat{\mathbf{y}} = \tilde{\mathbf{C}}_m(\mathbf{w}, \hat{\mathbf{u}}_z) \tag{137}$$

From Equation (115)

$$\mathbf{w} = \tilde{\mathbf{N}}(\mathbf{z}) \tag{138}$$

Taking the time derivative and applying the chain rule to (138) gives

$$\dot{\mathbf{w}} = \frac{\partial \tilde{\mathbf{N}}(\mathbf{z})}{\partial \mathbf{z}}\dot{\mathbf{z}} \tag{139}$$

where $\partial \tilde{\mathbf{N}}(\mathbf{z})/\partial \mathbf{z}$ is the Jacobian of $\tilde{\mathbf{N}}(\mathbf{z})$ with respect to $\mathbf{z}$. Substituting (138) and (139) into the modal equations of motion and premultiplying by $(\partial \tilde{\mathbf{N}}(\mathbf{z})/\partial \mathbf{z})^{-1}$ yields the state equations in physical coordinates as

$$\dot{\mathbf{z}} = \left(\frac{\partial \tilde{\mathbf{N}}(\mathbf{z})}{\partial \mathbf{z}}\right)^{-1} \tilde{\mathbf{A}}_m(\tilde{\mathbf{N}}(\mathbf{z})) + \left(\frac{\partial \tilde{\mathbf{N}}(\mathbf{z})}{\partial \mathbf{z}}\right)^{-1} \hat{\mathbf{u}}_m(\tilde{\mathbf{N}}(\mathbf{z})) \tag{140}$$
$$= \tilde{\mathbf{A}}'(\mathbf{z}) + B'(\mathbf{z})\hat{\mathbf{u}}_z$$

and

$$\hat{\mathbf{y}} = \tilde{\mathbf{C}}_m(\tilde{\mathbf{N}}(\mathbf{z}), \hat{\mathbf{u}}_z) \tag{141}$$

Here the prime symbol represents the possibility that the transformation does not allow the true equations of motion to be recovered in transforming them from modal to physical coordinates. For example, if a power series approximation for the modes is used, as suggested by Shaw and Pierre, then for practical purposes, the series must be truncated for most problems. Since the modes are an approximation of the true modes, only an approximation of the true modal equations will be found. Here we have not distinguished between the "true" modal equations and the approximation to them since in this work they will always be derived from the equations of motion and thus will in general be approximate, the exception being similar normal modal systems. Since the equations of motion are derived in physical space and they can also be found again by the transformation from modal coordinate to physical coordinates it is important to point out that the physical space equations of motion derived from the modal equations will not be as correct as when derived from first principles. However, the backwards transformation (from modal to physical coordinates) can be useful in determining how accurate the transformations are.

3.7.2. Modal Controllability of a Nonlinear Normal Modal System

As for linear controllability, the *ideal controllability* case is where each modal equation can be forced individually. If the force vector $\hat{\mathbf{u}}_z$ can be written

$$\hat{\mathbf{u}}_z = P_{um}(\mathbf{w})\hat{\mathbf{u}}_{m_{\text{even}}} \tag{142}$$

where $\hat{\mathbf{u}}_{m_{\text{even}}}$ is an $N \times 1$ vector of independent modal forces and $P_{um}(\mathbf{w})$ is an $N \times N$ matrix defined as

$$P_{um}(\mathbf{w}) = B(\tilde{\mathbf{M}}(\mathbf{w}))_{\text{even}}^{-1} \left(\frac{\partial \tilde{\mathbf{M}}(\mathbf{w})}{\partial \mathbf{w}} \right)_{\text{even}} \tag{143}$$

then the nonlinear normal modes are individually controllable. In order for Equation (143) to have a solution, the matrix $\tilde{B}(\tilde{\mathbf{M}}(\mathbf{w}))_{\text{even}}$ must be non-singular. Since this criteria will seldom be met, a more useful way to look at the controllability of a system is to look at controllability norms similar to those defined in Section 3.5.3.

Just as in linear systems, it is difficult to obtain a useful meaning from a controllability norm if the inputs are not normalized. Thus we define a matrix P_n for the nonlinear system identical to the linear system such that

$$P_{n_{i,i}} = \frac{1}{\max(\hat{u}_i)} \tag{144}$$

and define $B_{mn}(\mathbf{w})$ and $\hat{\mathbf{u}}_n$ such that

$$\left(\frac{\partial\tilde{\mathbf{M}}(\mathbf{w})}{\partial\mathbf{w}}\right)^{-1} B(\tilde{\mathbf{M}}(\mathbf{w}))\hat{\mathbf{u}}_z = \left[\left(\frac{\partial\tilde{\mathbf{M}}(\mathbf{w})}{\partial\mathbf{w}}\right)^{-1} B(\tilde{\mathbf{M}}(\mathbf{w}))P_n^{-1}\right][P_n\hat{\mathbf{u}}_z]$$

$$= [B_{mn}(\mathbf{w})][\hat{\mathbf{u}}_{zn}] \tag{145}$$

where $\mathbf{u}_n$ represents the normalized force vector and $B_{mn}(\mathbf{w})$ represents the modal "B" matrix. The element $B_{mn_{q,r}}(\mathbf{w})$ represents the scaling of the rth input force to the qth mode. Note that since we have changed the ordering of the elements of the state vector for the nonlinear system (relative to the ordering of linear system modal equations in Section 3.5), $B_{mn_{q,r}}(\mathbf{w})$ represents the element of $B_{mn}(\mathbf{w})$ in the $2q$th row and rth column. Thus we define

$$\mathbb{C}_{q,r} = |B_{mn_{q,r}}(\mathbf{w})| \tag{146}$$

to be the controllability norm of the qth mode from the rth input. This norm also represents the maximum force which may be applied to the rth mode from the qth input. It is also possible to define the controllability norm of the qth mode from set R of inputs. This is given by

$$\mathbb{C}_{q,R} = (B_{mn_{q,R}}(\mathbf{w})B_{mn_{q,R}}^T(\mathbf{w}))^{1/2} \tag{147}$$

For the case of repeated modes, the matrix $B_{mn_{q,R}}$ can be defined to be the Rth columns and the $2q$th rows (corresponding to the qth mode), of the matrix $B_{mn}(\mathbf{w})$. The modal controllability norm is then given by

$$\mathbb{C}_{q,R} = \det(B_{mn_{q,R}}(\mathbf{w})B_{mn_{q,R}}^T(\mathbf{w}))^{1/2N_q} \tag{148}$$

where N_q is the multiplicity of the qth mode. Note that this definition breaks down in the absence of pure modal motion and thus becomes an approximation since the modal equations are not completely decoupled. For a linear system, these results reduce to the results of Section 3.5.3.

3.7.3. Modal Observability of a Nonlinear Normal Systems

A system is said to be ideally observable if from the outputs of the system, all of the states of the system can be determined. Thus we define the inverse function

$$\mathbf{w} = \tilde{\mathbf{Q}}(\hat{\mathbf{y}}, \hat{\mathbf{u}}_z) \tag{149}$$

such that

$$\hat{\mathbf{y}} = \hat{\mathbf{C}}_m(\tilde{\mathbf{Q}}(\hat{\mathbf{y}}, \hat{\mathbf{u}}_z), \hat{\mathbf{u}}_z) \tag{150}$$

If a function $\tilde{\mathbf{Q}}(\hat{\mathbf{y}}, \hat{\mathbf{u}}_z)$ can be found which satisfies (150) then the system is said to have "ideal observability." Since this will not be possible in the majority of systems it is useful to quantify how observable each mode is from each actuator. In Section 3.5.3 the modal controllability for a linear system was defined based on the sensitivity of $\mathbf{y}$ to the modal vector $\mathbf{w}$ after scaling the output equation with respect to the sensitivities of the sensors. The same arguments for scaling the linear output equation hold for the nonlinear output equation. Thus, the output Equation (132) becomes

$$\mathbf{y}_n = \tilde{\mathbf{C}}_{mn}(\mathbf{w}, \hat{\mathbf{u}}_z) \tag{151}$$

where $\hat{\mathbf{y}}_n = P_{ncn}^{-1}\hat{\mathbf{y}}$ $\tilde{\mathbf{C}}_{mn}(\mathbf{w}, \hat{\mathbf{u}}_z) = P_{ncn}^{-1}\tilde{\mathbf{C}}_n(\mathbf{w}, \hat{\mathbf{u}}_z)$ and P_{ncn}^{-1} is a diagonal matrix consisting of the sensitivities of the sensors. The sensitivities of $\hat{\mathbf{y}}_n$ to $\mathbf{w}$ is then the Jacobian of $\hat{\mathbf{y}}_n$ with respect to $\mathbf{w}$

$$S(\mathbf{w}, \hat{\mathbf{u}}_z) = \frac{\partial \hat{\mathbf{y}}_n}{\partial \mathbf{w}} = \frac{\partial \hat{\mathbf{C}}_{mn}(\mathbf{w}, \hat{\mathbf{u}}_z)}{\partial \mathbf{w}} \tag{152}$$

The observability of the ith degree of freedom the rth sensor may then be defined as

$$\varnothing_{r,i} = |S_{r,i}(\mathbf{w}, \hat{\mathbf{u}}_z)| \tag{153}$$

as in the linear system. This definition, however, neglects the detrimental effect of a sensor from which the readings are an even function of a mode. For example, an accelerometer sensing radial acceleration of a rotating shaft can sense rotation but cannot determine the direction of the rotation. Depending on what the purpose of the output is, a more useful norm might be

$$\varnothing_{\text{odd}_{r,i}} = |\$_{r,i}(\mathbf{w}, \hat{\mathbf{u}}_z)| \tag{154}$$

where

$$\$(\mathbf{w}, \hat{\mathbf{u}}_z) = \frac{1}{2}(S(\mathbf{w}, \hat{\mathbf{u}}_z) + S(-\mathbf{w}, \hat{\mathbf{u}}_z)) \tag{155}$$

Defining the norm in this fashion causes only the odd part of the output to be considered in the observability norm. A gross measure of observability from the sensors R can then be defined as

$$\varnothing_{R,i} = (\$_{R,i}(\mathbf{w}, \hat{\mathbf{u}}_z)^T \$_{R,i}(\mathbf{w}, \hat{\mathbf{u}}_z))^{1/2} \tag{156}$$

where $\$_{R,i}(\mathbf{w}, \hat{\mathbf{u}}_z)$ is a vector containing the Rth elements of the ith column of $\$(\mathbf{w}, \hat{\mathbf{u}}_z)$.

3.8. NONLINEAR MODAL CONTROL DESIGN

In this section the methods for controlling a nonlinear modal system under various circumstances are discussed. In Section 3.4 linearizing control was discussed as well as the requirements for applying linearizing control. Under the conditions of ideal observability and ideal controllability, and also in some cases with non-ideal controllability, linearizing control can be applied. However, Section 3.8.1 discusses a method for bypassing the linearizing control step to derive a controller which will yield the desired modal characteristics for systems with ideal controllability and ideal observability. Section 3.8.2 discusses the case of non-ideal observability. In most cases where the controllability is non-ideal, linearizing control cannot be applied. Section 3.8.3 discusses an approximate method for obtaining the desired modal characteristics for such systems.

3.8.1. Ideal Case - Ideal Controllability and Ideal Observability

The simplest case to design control for is the case of ideal controllability and ideal observability as discussed in Sections 3.7.2 and 3.7.3. Often in this case linearizing control is applied to the system. Although this approach is quite simple, it has at least two major drawbacks. The first is that unless the model of the structure is perfect, the structure with linear control will not be linear either. In systems exhibiting nonlinear normal modes, control laws can be designed in the nonlinear modal space, thus bypassing the need for linearizing control. The second is that linear control is not the optimal control in general. This can be seen by comparing the optimal open loop control to the best closed loop control. Generally, they do not match. Linear control is simply the easiest control to design and implement. Thus what is usually called optimal control is usually the optimal linear control. A method for applying nonlinear modal control for the special case of ideal controllability and ideal observability is demonstrated below.

The nonlinear state space equations in the physical coordinate space are given by Equations (131) and (132) as

$$\dot{\mathbf{z}} = \tilde{\mathbf{A}}(\mathbf{z}) + B(\mathbf{z})\hat{\mathbf{u}}_z = A(\mathbf{z})\mathbf{z} + B(\mathbf{z})\hat{\mathbf{u}}_z \tag{157}$$

$$\hat{\mathbf{y}} = \tilde{\mathbf{C}}(\mathbf{z}, \hat{\mathbf{u}}_z) \tag{158}$$

and in modal coordinates by Equations (134) and (135) as

$$\dot{\mathbf{w}} = \left(\frac{\partial \tilde{\mathbf{M}}(\mathbf{w})}{\partial \mathbf{w}}\right) A(\tilde{\mathbf{M}}(\mathbf{w}))\tilde{\mathbf{M}}(\mathbf{w}) + \left(\frac{\partial \tilde{\mathbf{M}}(\mathbf{w})}{\partial \mathbf{w}}\right)^{-1} B(\tilde{\mathbf{M}}(\mathbf{w}))\hat{\mathbf{u}}_z \tag{159}$$

$$= \tilde{\mathbf{A}}_m(\mathbf{w}) + \hat{\mathbf{u}}_m(\mathbf{w}, \hat{\mathbf{u}}_z)$$

$$\hat{\mathbf{y}} = \tilde{\mathbf{C}}(\tilde{\mathbf{M}}(\mathbf{w}), \hat{\mathbf{u}}_z) = \tilde{\mathbf{C}}_m(\mathbf{w}, \hat{\mathbf{u}}_z) \tag{160}$$

The desired modal state equation is given as

$$\dot{\mathbf{w}} = \tilde{\mathbf{A}}_{md}(\mathbf{w}) \tag{161}$$

where $\tilde{\mathbf{A}}_{md}(\mathbf{w})$ is the desired modal function. The objective of the control is to chose a control force

$$\hat{\mathbf{u}}_m = \mathbf{G}_m(\hat{\mathbf{y}}) \tag{162}$$

such that the desired behavior is achieved. Since ideal controllability means that each mode can be controlled independently of the others, this is possible. Combining Equations (161), (162) and (159) yields

$$\tilde{\mathbf{A}}_{md_{\text{even}}}(\mathbf{w}) = \tilde{\mathbf{A}}_{m_{\text{even}}}(\mathbf{w}) + \mathbf{G}_{m_{\text{even}}}(\mathbf{w}) \tag{163}$$

where the subscript even means that we are considering only the even elements of the vectors. Remember from Section 6 (see Equations (90) and (91)) that the odd numbered state equations are simply identity statements. Thus the control law in terms of the modal coordinates is

$$\mathbf{G}_{m_{\text{even}}}(\mathbf{w}) = \tilde{\mathbf{A}}_{md_{\text{even}}}(\mathbf{w}) - \tilde{\mathbf{A}}_{m_{\text{even}}}(\mathbf{w}) \tag{164}$$

The constraint for ideal observability is given in Equations (149) and (150) as

$$\mathbf{w} = \tilde{\mathbf{Q}}(\hat{\mathbf{y}}, \hat{\mathbf{u}}_z) \tag{165}$$

such that

$$\hat{\mathbf{y}} = \tilde{\mathbf{C}}_m(\tilde{\mathbf{Q}}(\hat{\mathbf{y}}, \hat{\mathbf{u}}_z), \hat{\mathbf{u}}_z) \tag{166}$$

Since we wish to find $\mathbf{w}$ as a function of $\hat{\mathbf{y}}$, the output equation must not be dependent on $\hat{\mathbf{u}}$. Thus it is required that

$$\mathbf{w} = \tilde{\mathbf{Q}}(\hat{\mathbf{y}}) \tag{167}$$

Substituting for $\mathbf{w}$ in Equation (164) using Equation (167) gives

$$\begin{aligned}
\mathbf{G}_{m_{\text{even}}}(\hat{\mathbf{y}}) &= \tilde{\mathbf{A}}_{md_{\text{even}}}(\tilde{\mathbf{Q}}(\hat{\mathbf{y}})) - \tilde{\mathbf{A}}_{m_{\text{even}}}(\tilde{\mathbf{Q}}(\hat{\mathbf{y}})) \\
&= \tilde{\mathbf{A}}_{md_{\text{even}}}(\hat{\mathbf{y}}) - \tilde{\mathbf{A}}_{m_{\text{even}}}(\hat{\mathbf{y}})
\end{aligned} \tag{168}$$

Using Equation (142) the control law in physical coordinates is

$$\mathbf{G}_{z_{even}} = B_{even}(\tilde{\mathbf{M}}(\mathbf{w}))^{-1} \left(\frac{\partial \tilde{\mathbf{M}}(\mathbf{w})}{\partial \mathbf{w}}\right)_{even} \mathbf{G}_{m_{even}}(\hat{\mathbf{y}}) \qquad (169)$$

Since we cannot feedback a control law which is a function of anything but the outputs, substituting Equation (167) for $\mathbf{w}$ gives the feedback control law in physical space as

$$\mathbf{G}_{z_{even}}(\hat{\mathbf{y}}) = B_{even}(\tilde{\mathbf{M}}(\tilde{\mathbf{Q}}(\hat{\mathbf{y}})))^{-1} \left(\frac{\partial \tilde{\mathbf{M}}(\mathbf{w})}{\partial \mathbf{w}}\right)_{even}\bigg|_{\mathbf{w}=\tilde{\mathbf{Q}}(\tilde{\mathbf{y}})} \mathbf{G}_{m_{even}}(\hat{\mathbf{y}}) \qquad (170)$$

where the control $\mathbf{G}_{m_{even}}(\hat{\mathbf{y}})$ law is given by Equation (168). Thus for the ideally observable and ideally controllable case, any desired modal control can be transformed into physical space and implemented to yield a closed loop system with the desired modal characteristics.

3.8.2. Non-Ideal Observability

In a system with non-ideal observability, it is not possible to determine the modal coordinates as functions of the output $\hat{\mathbf{y}}$. Thus, in deriving a control law, it is necessary to derive the control law such that $\hat{\mathbf{y}}$ can be substituted for the modal coordinates $\mathbf{w}$. Then, if the transformation of Equation (149) can be performed, the control law can be implemented in physical space. This is possible for systems exhibiting similar normal modes and may be possible for some systems exhibiting non-similar normal modes. In general, however, transformation of modal control laws for systems with non-similar normal modes will not be exactly possible because $(\partial \tilde{\mathbf{M}}(\mathbf{w})/\partial \mathbf{w})$ cannot be written in terms of $\hat{\mathbf{y}}$ unless the system is ideally observable. In these cases it is necessary to approximate $(\partial \tilde{\mathbf{M}}(\mathbf{w})/\partial \mathbf{w})$ by using a linear (or low amplitude) approximation for $(\partial \tilde{\mathbf{M}}(\mathbf{w})/\partial \mathbf{w})$. When this is necessary, a useful check is to transform the physical coordinate vector into modal coordinates using

$$\mathbf{G}_m = \left(\frac{\partial \tilde{\mathbf{M}}(\mathbf{w})}{\partial \mathbf{w}}\right)^{-1} B(\tilde{\mathbf{M}}(\mathbf{w}))\mathbf{G}_z \qquad (171)$$

to examine how much degradation has taken place in the control law due to the transformation from modal to physical coordinates. Note that all of the coordinates are used in the transformation since the transformation of the control law to modal space most likely will effect the odd numbered modal state Equations (the identity statements) as well as the even numbered modal state equations.

3.8.3. Non-Ideal Controllability

A more realistic case is the case of ideal observability, but non-ideal controllability. With sensors becoming more cost effective and less intrusive to the system every day, more and more structures are capable of being outfitted fully with sensors (such as smart structures). The increased power and cost-effectiveness of controller computers also makes the processing of large amounts of data signals more feasible. On the other hand, it is generally unrealistic to distribute actuators throughout an entire structure due to either cost, weight constraints, or energy constraints. For this case, the approach recommended is to find $\mathbf{u}_m$ in terms of the modal coordinates and $\mathbf{u}_z$ (see Equation (159)). For this transformation, constraints on the elements of $\mathbf{u}_m$ will become apparent in terms of $\mathbf{u}_z$ and the modal coordinates. These constraints can then be written in terms of relationships between the elements of $\mathbf{u}_m$. Designing the modal control law using these constraints will allow the control law to be feasible when transformed into the physical coordinate control law. In order to find the control law in physical space, a modified version of Equation (170) is used such that

$$B_{\text{even}}(\tilde{\mathbf{M}}(\mathbf{w}))G_{z_{\text{even}}}(\hat{\mathbf{y}}) = \left(\frac{\partial \tilde{\mathbf{M}}(\mathbf{w})}{\partial \mathbf{w}}\right)_{\text{even}} \Bigg|_{\mathbf{w}=\tilde{\mathbf{Q}}(\hat{\mathbf{y}})} G_{m_{\text{even}}}(\hat{\mathbf{y}}) \qquad (172)$$

where the control $\mathbf{G}_{m_{\text{even}}}(\hat{\mathbf{y}})$ law is given by Equation (168). Since $B_{\text{even}}(\tilde{\mathbf{M}}(\mathbf{w}))$ is not invertible, it must then be factored out of the right side of Equation (172) to yield $\mathbf{G}_{z_{\text{even}}}(\hat{\mathbf{y}})$.

Examples 1 and 3 illustrate this concept.

3.8.4. Some Conclusions About Nonlinear Control

Even if a system is not ideally controllable or ideally observable, a control law can be derived intuitively in certain cases by assuming that the coupling caused by the nonlinear control law in modal space is negligible. The procedure is as follows. First a feasible control law is derived in modal space based on the concepts of the previous two sections. This control law is then transformed into physical coordinates using an approximate transformation if necessary. For instance, this may require using the linearized coordinate transformation because some of the modal states of the system are not observable. Thus a feasible control law has been designed and transformed into physical space. The next step is to compile the new function vector $\hat{\mathbf{A}}(\mathbf{z})$ by combining the control law with the existing vector $\tilde{\mathbf{A}}(\mathbf{z})$. Nonlinear modal analysis can then be applied to the "new" system (The closed loop system). The new modal equations of the closed loop system and the new nonlinear mode shapes for the closed loop system can then be examined to see if the control law did achieve the desired effect. If not, this approach can be repeated until a control law which yields the desired effect has been designed. This is analogous to performing an eigenanalysis on the closed loop system for a linear system in order to observe the effectiveness of the controller.

This step can also be useful for nonlinear systems for which linear control was derived based on a linearized model of the system. Analyzing the modal characteristics of the closed loop nonlinear system with linear feedback will give a much better prediction of the modal properties then using the linearized model to predict modal properties.

3.9. EXAMPLES

In this section, we consider three example problems. The first is an ideally controllable and ideally observable linear two degree of freedom system. The observability of the first degree of freedom (the displacement of x_1) is found and a comparison of the gross observability norm of Hughes and Skelton[10] is made with the observability norms proposed in Section 3.5. An example modal control law is then derived for the same system but with non-ideal controllability.

The second problem is an ideally observable and ideally controllable nonlinear similar modal system. Control is applied to add damping to the system and decouple the nonlinear modal equations. Thus the global response characteristics can be derived from the characteristics of the modal equations.

The third example is a nonlinear non-similar modal system which is ideally observable but not ideally controllable. A control law is derived to yield linear modal damping at low amplitudes but much greater damping at high amplitudes of vibration. From observation it can be seen that example 1 is the linearized modal for example 3. Thus the comparison is made between the linear control derived using the linear model and the nonlinear modal control derived using the nonlinear modal equation by applying the control to the "true" nonlinear system. Simulations show that the desired effect has been achieved.

3.9.1. Example 1: A Linear System With Non-Ideal Controllability and Ideal Observability

Consider the following linear system where

$$V_i = \frac{1}{2}(x_i - x_{i-1})^2 \tag{173}$$

Note that $x_0 = z_4 = 0$. The equations of motion are

$$\begin{bmatrix} \dot{x}_1 \\ \dot{y}_1 \\ \dot{x}_2 \\ \dot{y}_2 \end{bmatrix} = \begin{bmatrix} 0 & 1 & 0 & 0 \\ -2 & 0 & 1 & 0 \\ 0 & 0 & 0 & 1 \\ 1 & 0 & -2 & 0 \end{bmatrix} \begin{bmatrix} x_1 \\ y_1 \\ x_2 \\ y_2 \end{bmatrix} + \begin{bmatrix} 0 & 0 \\ 1 & 0 \\ 0 & 0 \\ 0 & 1 \end{bmatrix} \begin{bmatrix} \hat{u}_1 \\ \hat{u}_2 \end{bmatrix} \tag{174}$$

$$\hat{\mathbf{y}} = I\mathbf{z} \tag{175}$$

$$\mathbf{z} = \mathbf{M}(\mathbf{w})\mathbf{w} = \begin{bmatrix} 1 & 0 & 1 & 0 \\ 0 & 1 & 0 & 1 \\ 1 & 0 & -1 & 0 \\ 1 & 1 & 0 & -1 \end{bmatrix} \begin{bmatrix} u_1 \\ v_1 \\ u_2 \\ v_2 \end{bmatrix} \tag{176}$$

From Equations (134) and (135)

$$\dot{\mathbf{w}} = A_m \mathbf{w} + \begin{bmatrix} 0 \\ .5(\hat{u}_1 + \hat{u}_2) \\ 0 \\ .5(\hat{u}_1 + \hat{u}_2) \end{bmatrix} \tag{177}$$

$$\hat{\mathbf{y}} = \mathbf{M}(\mathbf{w})\mathbf{w} \tag{178}$$

where

$$A_m = \begin{bmatrix} 0 & 1 & 0 & 0 \\ -1 & 0 & 0 & 0 \\ 0 & 0 & 0 & 1 \\ 0 & 0 & -3 & 0 \end{bmatrix} \tag{179}$$

From Equation (152), the sensitivity of degree of freedom #1 from all sensors is

$$\$_{[1,2,3,4],1} = \begin{bmatrix} 1 \\ 0 \\ 1 \\ 0 \end{bmatrix} \tag{180}$$

and the gross measure of observability of degree of freedom #1 from all of the sensors is, from Equation (156),

$$\phi_{[1,2,3,4],1} = \sqrt{2} \tag{181}$$

In fact, the gross measure of observability of all degrees of freedom is $\sqrt{2}$ when using all sensors. If we consider the system with the displacement sensors removed, the gross measures of observability for the modal displacements are zero. If we instead were to use the gross modal observability norm of Equation (25) of Hughes and Skelton[10], we would find a nonzero modal observability norm for each of the modes. Using the notation of Hughes and Skelton,

$$\hat{P} = PM_{\text{odd}} = IM_{\text{odd}} \qquad \hat{R} = RM_{\text{even}} = IM_{\text{even}} \tag{182}$$

Thus

$$P_1 = \begin{bmatrix} 1 \\ 1 \end{bmatrix}, \qquad R_1 = \begin{bmatrix} 1 \\ 1 \end{bmatrix} \tag{183}$$

and the negative square of the first natural frequency is $k = -1$. The matrix F_q is then

$$F_q = \begin{bmatrix} P_1 & R_1 \\ -kR_1 & P_1 \end{bmatrix} = \begin{bmatrix} 1 & 1 \\ 1 & 1 \\ -1 & 1 \\ -1 & 1 \end{bmatrix} \tag{184}$$

and the observability norm for the first mode is then given by

$$\varnothing = \det(F_q^T f_q)^{1/4} = 2 \tag{185}$$

This happens even though there is no way to directly observe the modal displacements (i.e. without an observer) and therefore they are not available for control. This measure of observability is not appropriate when designing a control law based on only the directly observable states.

Consider the case where no actuator exists at x_1, thus $\hat{u}_1 = 0$. In attempting to design a control law such that $t_s = 60$ sec the settling time is given as

$$t_s = \frac{3}{\xi \omega} = 60 \tag{186}$$

such that

$$2\xi\omega = \frac{6}{60} = .1 \tag{187}$$

Thus we would want to feedback $0.1v$ for each mode. However, due to the constraint that $u_1 = 0$, any control applied to one mode must be the negative of the control applied to the other. Thus we choose

$$\hat{u}_2 = .2(v_2 - v_1) = [0 \ - .2 \ 0 \ .2]\mathbf{w} \tag{188}$$

such that the modal matrix $A_m(\mathbf{w})$ is

$$A_m(\mathbf{w}) = \begin{bmatrix} 0 & 1 & 0 & 0 \\ -1 & -.1 & 0 & +.1 \\ 0 & 0 & 0 & 1 \\ 0 & .1 & -3 & -.1 \end{bmatrix} \tag{189}$$

after the control law

$$G_m = \begin{bmatrix} 0 & 0 & 0 & 0 \\ 0 & -.1 & 0 & .1 \\ 0 & 0 & 0 & 0 \\ 0 & .1 & 0 & -.1 \end{bmatrix} \mathbf{w} \tag{190}$$

is applied. Substituting $\mathbf{w} = M^{-1}\hat{\mathbf{y}}$ the modal control law in modal coordinates written in terms of the output $\hat{\mathbf{y}}$ is then

$$G_m = \begin{bmatrix} 0 & 0 & 0 & 0 \\ 0 & 0 & 0 & -.1 \\ 0 & 0 & 0 & 0 \\ 0 & 0 & 0 & .1 \end{bmatrix} \begin{bmatrix} \hat{y}_1 \\ \hat{y}_2 \\ \hat{y}_3 \\ \hat{y}_4 \end{bmatrix} \tag{191}$$

From Equation (172)

$$B(\mathbf{z})G_z(\hat{\mathbf{y}}) = M(\mathbf{w})G_m(\hat{\mathbf{y}}) = \begin{bmatrix} 0 & 0 & 0 & 0 \\ 0 & 0 & 0 & 0 \\ 0 & 0 & 0 & 0 \\ 0 & 0 & 0 & -.2 \end{bmatrix} \hat{\mathbf{y}} \tag{192}$$

The control $\mathbf{G}_m(\hat{\mathbf{y}})$ can then be found to be

$$G_z(\hat{\mathbf{y}}) = \begin{bmatrix} 0 & 0 & 0 & 0 \\ 0 & 0 & 0 & -.2 \end{bmatrix} \hat{\mathbf{y}} \tag{193}$$

Substituting $\mathbf{G}_z(\hat{\mathbf{y}})$ for $\mathbf{u}_z$ and $\hat{\mathbf{y}} = I\mathbf{z}$ yields the closed loop state equation as

$$\dot{\mathbf{z}} = \begin{bmatrix} 0 & 1 & 0 & 0 \\ -2 & 0 & 1 & 0 \\ 0 & 0 & 0 & 1 \\ 1 & 0 & -2 & -.2 \end{bmatrix} \mathbf{z} \tag{194}$$

Modal analysis of the closed loop equation shows that the settling time for the first mode is 59.7 seconds and the settling time for the second mode is 60.3 seconds. An iterative analysis shows that increasing the gain by 0.5% will yield a settling time of 60 seconds or less for both modes.

3.9.2. Example 2: A Nonlinear Similar Modal System With Ideal Controllability and Ideal Observability

Consider the same system as example 1 but with cubic springs such that

$$V_i = \frac{1}{4}(x_i - x_{i-1})^4 \tag{195}$$

The equations of motion for the system are

$$\begin{bmatrix} \dot{x}_1 \\ \dot{y}_1 \\ \dot{x}_2 \\ \dot{y}_2 \end{bmatrix} = \begin{bmatrix} y_1 \\ -2x_1^3 + x_2^3 \\ y_2 \\ -2x_2^3 + x_1^3 \end{bmatrix} + \begin{bmatrix} 0 & 0 \\ 1 & 0 \\ 0 & 0 \\ 0 & 1 \end{bmatrix} \hat{\mathbf{u}}_2 \tag{196}$$

Applying the method of Shaw and Pierre, the nonlinear similar modes are found to be (See Appendix C)

mode 1:

$$u_2 = u_1, \qquad v_2 = v_1 \qquad (197)$$

mode 2:

$$u_2 = -u_1, \qquad v_2 = -v_1 \qquad (198)$$

The modal coordinate transformation is then

$$\mathbf{z} = \begin{bmatrix} 1 & 0 & 1 & 0 \\ 0 & 1 & 0 & 1 \\ 0 & 0 & -1 & 0 \\ 0 & 1 & 0 & -1 \end{bmatrix} \mathbf{w} \qquad (199)$$

Transforming the equations of motion into modal coordinates gives (See Appendix E)

$$\begin{bmatrix} \dot{u}_1 \\ \dot{v}_1 \\ \dot{u}_2 \\ \dot{v}_2 \end{bmatrix} = \begin{bmatrix} v_1 \\ -u_1^3 - 3u_1u^2 - 2 \\ v_2 \\ -3u_2^3 - 9u_2u_1^2 \end{bmatrix} + \begin{bmatrix} 0 \\ .5(\hat{u}_1 + \hat{u}_2) \\ 0 \\ .5(\hat{u}_1 + \hat{u}_2) \end{bmatrix} \qquad (200)$$

$$\hat{\mathbf{y}} = \begin{bmatrix} 1 & 0 & 1 & 0 \\ 0 & 1 & 0 & 1 \\ 1 & 0 & -1 & 0 \\ 0 & 1 & 0 & -1 \end{bmatrix} \begin{bmatrix} u_1 \\ v_1 \\ u_2 \\ v_2 \end{bmatrix} \qquad (201)$$

Since the vector $\mathbf{w}$ can be found as a function of $\hat{\mathbf{y}}$ the system is perfectly observable. Also, since

$$B_{m_{even}} = \frac{1}{2}\begin{bmatrix} 1 & 1 \\ 1 & -1 \end{bmatrix} \qquad (202)$$

is non-singular, the system is ideally controllable as well.

Because the system is ideally controllable and ideally observable, any desired control law can be designed. For instance, a control law can be designed to decouple the modal equations and add damping as well. Applying a decoupling control law allows the determination of global stability in terms of the stability of each mode. Applying an energy dissipation term to each mode stabilizes each mode, thus making the global system stable. Take the desired modal equations to be

$$\hat{A}_{md} = \begin{bmatrix} v_1 \\ -u_1^3 - u_1^2 v_1 \\ v_2 \\ -3u_2^3 - u_2^2 v_2 \end{bmatrix} \qquad (203)$$

The control law in terms of the modal coordinates is then

$$G_m(\mathbf{w}) = \begin{bmatrix} 0 \\ 2u_1u_2^2 - u_1^2v_1 \\ 0 \\ 9u_2u_1^2 - u_2^2v_2 \end{bmatrix} \tag{204}$$

Substituting $\mathbf{w} = \hat{\mathbf{Q}}(\hat{\mathbf{y}}) = M^{-1}(\hat{\mathbf{y}})$ for $\mathbf{w}$ yields the modal control law in terms of the outputs as (See Appendix F)

$$G_m(\hat{\mathbf{y}}) = \begin{bmatrix} 0 \\ \dfrac{3(\hat{y}_1 - \hat{y}_3)^2(\hat{y}_1 + \hat{y}_3)}{8} - \dfrac{(\hat{y}_1 + \hat{y}_3)^2(\hat{y}_2 + \hat{y}_1)}{8} \\ 0 \\ \hat{y}_1^3 + \dfrac{3\hat{y}_1^2\hat{y}_3}{2} - \dfrac{3\hat{y}_1\hat{y}_3^2}{2} - \hat{y}_3^3 \end{bmatrix} \tag{205}$$

Applying Equation (169) the control law in physical coordinates is

$$G_z(\hat{\mathbf{y}})_{\text{even}} = \begin{bmatrix} h_1 + h_2 \\ h_1 - h_2 \end{bmatrix} \tag{206}$$

where

$$h_1 = \frac{3(\hat{y}_1 - \hat{y}_3)^2(\hat{y}_1 + \hat{y}_3)}{8} - \frac{3(\hat{y}_1 + \hat{y}_3)^2(\hat{y}_2 + \hat{y}_1)}{8} \tag{207}$$

and

$$h_2 = \hat{y}_1^3 + \frac{3\hat{y}_1^2\hat{y}_3}{2} - \frac{3\hat{y}_1\hat{y}_3^2}{2} - \hat{y}_3^3 \tag{208}$$

3.9.3. Example 3: A Nonlinear Nonsimilar Modal System With Non-Ideal Controllability and Ideal Observability

Consider the same system of Figure 4 but with springs such that potential energies are

$$V_1 = \frac{1}{2}x_1^2 + \frac{1}{8}x_1^4$$
$$V_2 = \frac{1}{2}(x_2 - x_1)^2 \tag{209}$$

and

$$V_3 = \frac{1}{2}x_2^2 \tag{210}$$

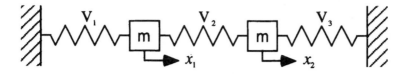

Figure 4. A two degree of freedom oscillator.

Also consider the case where only the second degree of freedom can be actuated. The equations of motion for the system are then

$$\begin{bmatrix} \dot{x}_1 \\ \dot{y}_1 \\ \dot{x}_2 \\ \dot{y}_2 \end{bmatrix} = \begin{bmatrix} y_1 \\ -2x_1 - .5x_1^3 + x_2 \\ y_2 \\ x_1 - 2x_2 \end{bmatrix} + \begin{bmatrix} 0 \\ 0 \\ 0 \\ 1 \end{bmatrix} \hat{u}_z \tag{211}$$

and

$$\hat{\mathbf{y}} = I\mathbf{z} \tag{212}$$

Using the method of Shaw and Pierre, the equations of motion in modal coordinates are

$$\begin{bmatrix} \dot{u}_1 \\ \dot{v}_1 \\ \dot{u}_2 \\ \dot{v}_2 \end{bmatrix} = \begin{bmatrix} v_1 \\ -u_1 - \dfrac{1}{3}u_1^3 + \dfrac{1}{4}u_1v_1^2 - \dfrac{3}{4}u_1^2u_2 - \dfrac{3}{4}u_1u_2^2 \\ v_2 \\ -3u_2 - \dfrac{4}{13}u_2^3 + \dfrac{3}{52}u_2v_2^2 - \dfrac{3}{4}u_1^2u_2 - \dfrac{3}{4}u_1u_2^2 \end{bmatrix} + \begin{bmatrix} \hat{u}_{m1} \\ \hat{u} + m2 \\ \hat{u}_{m3} \\ \hat{u}_{m4} \end{bmatrix} \tag{213}$$

where the modal control forces are

$$\hat{u}_{m1} = -\hat{u}_{m3} = -0.125u_1v_1 + 0.0288u_2v_2 \tag{214}$$

and,

$$\hat{u}_{m2} = -\hat{u}_{m4} = 0.500 + 0.0577u_2^2 - 0.188v_1^2 + 0.0433v_2^2 \tag{215}$$

The output equation is then

$$\hat{\mathbf{y}} = \hat{\mathbf{M}}(\mathbf{w}) = \begin{bmatrix} u_1 + u_2 \\ v_1 + v_2 \\ u_1 - u_2 + .1667u_1^3 + 1.923u_2^3 + .25u_1v_1^2 + .05769u_2v_2^2 \\ v_1 - v_2 + .25v_1^3 + .05769v_2^3 + .2308u_2^2v_2 \end{bmatrix} \tag{216}$$

The mode shapes are given by

Mode 1

$$\begin{bmatrix} x_1 \\ y_1 \\ x_2 \\ y_2 \end{bmatrix} = \tilde{M}(w)\Bigg|_{u_2=0,v_2=0} = \begin{bmatrix} u_2 \\ v_1 \\ u_1 + \frac{1}{6}u_1^3 + .25u_1v_1^2 \\ v_1 + .25v_1^3 \end{bmatrix} \tag{217}$$

and

Mode 2

$$\begin{bmatrix} x_1 \\ y_1 \\ x_2 \\ y_2 \end{bmatrix} = \tilde{M}(w)\Bigg|_{u_1=0,v_1=0} = \begin{bmatrix} u_2 \\ v_2 \\ -u_2 + .1923u_2^3 + .05769u_2v_2^2 \\ -v_2 + .05769v_2^3 + .2308u_2^2v_2 \end{bmatrix} \tag{218}$$

Figure 5 shows mode shapes (displacement one versus displacement two) for various modal amplitudes of mode one.[30] Note that even the linearization of these modal motions are different depending on the modal amplitude. Figure 6 shows the mode shapes of mode two for various modal amplitudes.

The system is ideally observable since

$$\tilde{Q}(\hat{y}) = \tilde{N}(\hat{y}) \tag{219}$$

Also, as long as the modal control law satisfies Equations (214) and (215) it will be physically realizable. The relationships between the modal control forces can be written as

$$\hat{u}_{m1} = f\hat{u}_{m2} = -\hat{u}_{m3} = -f\hat{u}_{m4} \tag{220}$$

where

$$f = -0.25u_1v_1 + 0.0577u_2v_2 \tag{221}$$

Since the purpose of this example is to show how a desired nonlinear modal control can be transformed to physical coordinates and applied to the system, one will be proposed using heuristic observations. Consider a system in which very small oscillations are considered unimportant but larger oscillations are considered to be detrimental. A damping term could then be introduced which has a small coefficient for small amplitudes and a large coefficient for larger amplitudes. Possible modal control laws could then be

$$\hat{u}_{m2} = -.1\dot{u}_1(1 + \dot{u}_1^2 + u_1^2) \tag{222}$$

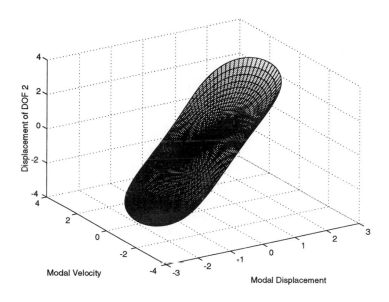

Figure 5. Mode 1.

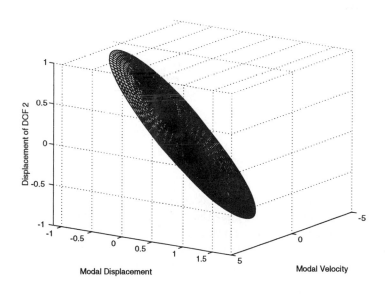

Figure 6. Mode 2.

and

$$\hat{u}_{m4} = -.1\dot{u}_2(1 + \dot{u}_2^2 + 3u_2^2) \tag{223}$$

However, because of the constraints of Equations (214) and (215), the following alternative is proposed

$$\hat{u}_{m2} = -.1\dot{u}_1(1 + \dot{u}_1^2 + u_1^2) + .1\dot{u}_2(1 + \dot{u}_2^2 + 3u_2^2) \tag{224}$$

and

$$\hat{u}_{m4} = -\hat{u}_{m2} \tag{225}$$

Thus the negative of the control law intended for mode 1 is applied to mode 2 and the negative of the control law intended for mode 2 is applied to mode 1 in addition to the desired controls. It is expected that since these cross terms are small, they should have little impact on the modal equations. Also, Nayfeh and Mook[17] show that for systems having cubic nonlinearities, the effect of the cross terms is negligible when no internal resonance exists.

The control law in modal coordinates is then

$$\begin{bmatrix} 0 \\ \hat{u}_{m2} \\ 0 \\ \hat{u}_{m4} \end{bmatrix} \tag{226}$$

Using Equation (169) to transform the control law to physical coordinates in terms of the output $\hat{\mathbf{y}}$ yields

$$B\hat{u}_z = \begin{bmatrix} 0 \\ 0 \\ 0 \\ \beta \end{bmatrix} \tag{227}$$

where

$$\begin{aligned}
\beta = & -0.2\hat{y}_4 + 0.0558\hat{y}_2(\hat{y}_1^2 + \hat{y}_3^2) - 0.212\hat{y}_1\hat{y}_2\hat{y}_3 + 0.00769\hat{y}_2^3 \\
& -0.1\hat{y}_4(\hat{y}_1^2 - \hat{y}_1\hat{y}_3 + \hat{y}_3^2) - 0.15\hat{y}_2^2\hat{y}_4 \\
& -0.0231\hat{y}_2\hat{y}_4^2 - 0.0596\hat{y}_2^3
\end{aligned} \tag{228}$$

From the equations of motion, it is clear that $B\hat{u}_z$ cannot have this form. All but the last element of $B\hat{u}_z$ must be identically zero. The error is believed to be from the approximate inverse (from modal to physical) coordinate transformation ($\tilde{\mathbf{N}}(\mathbf{z})$) as discussed in Section 3.6.4. Since the fourth element of $B\hat{u}_z$ should be equation to $\hat{u}_z$, $\hat{u}_z$ is taken to be

$$\begin{aligned}
\hat{u}_z = & -.2\hat{y}_2 - .01\hat{y}_4(\hat{y}_1^2 + \hat{y}_3^2) + .0442308\hat{y}_4(\hat{y}_1^2 + \hat{y}_3^2) - .188462\hat{y}_1\hat{y}_3\hat{y}_2 \\
& + .1\hat{y}_1\hat{y}_3\hat{y}_4 - .15\hat{y}_2^2\hat{y}_4 + .0230769\hat{y}_2\hat{y}_4^2 - .0403846\hat{y}_4^3
\end{aligned} \tag{229}$$

Comparing the equations of motion for this example with the equations of motion for example 1 shows that the system of example 1 is the linearized system of example 3. Furthermore, the linear modal control of example 1 is the linear part of the modal control derived for this example. Thus, example 1 is an example of the control law that could have been derived for this system if we derived a linear modal control law based on a linearized model. Therefore, the control law of example 1 has been applied to the nonlinear equations of motion for this example for comparison with the nonlinear modal control. Simulations of the controlled steady state response and transient response are shown in the following plots. The displacement plotted is that of x_1. Figures 7 and 8 show the transient response from an initial condition of $x_1 = 1$ and all other states equal to zero. It is clear that the control forces are greater at higher amplitudes for the nonlinear control then for the linear control. Also, the motion of the first mass is reduced much more quickly using the nonlinear modal control than the linear control. Figures 9 and 10 show the steady state results for a 1 rad/sec sinusoidal force on mass 2 with an amplitude of 0.01 and zero initial conditions. Clearly, the goal of having only a linear energy dissipation at low amplitudes has been met. Figures 11 and 12 show the steady state results for a rad/sec sinusoidal force on mass 2 with an amplitude of 1. Here the effect of the nonlinear control is clear. At higher amplitudes, the nonlinear control provides additional damping, beyond that of the linear control, and therefore reduces the steady state response. Thus this example shows two major points. The first is that, depending on system requirements, nonlinear control is usually better than linear control. This fact can be observed in most systems when finding the optimal open loop control. Generally, the optimal open loop control does not match the capability of any reasonably sized (low order) linear controller. The second is that modal control laws, be they linear or nonlinear, can be successfully derived for nonlinear modal systems even if the modes are non-similar.

3.10. CONCLUSIONS

A new method for controlling structures which exhibit nonlinear normal modes has been presented. Modal norms originally proposed for linear modal systems have been expanded for nonlinear modal systems. Depending on the extent of the nonlinearities and the amplitude of the oscillations, observability and controllability norms can vary considerably. These norms can be of considerable use even when considering other control methods. Unlike existing robust linear control methodologies, nonlinear modal control allows the consideration of the nonlinearities explicitly instead of as uncertainties. Yet, like linear modal control, it is a very intuitive approach which includes

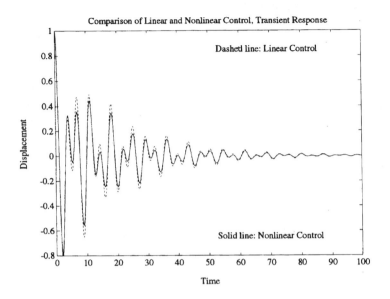

Figure 7. Transient response, displacement of x_1 verses time.

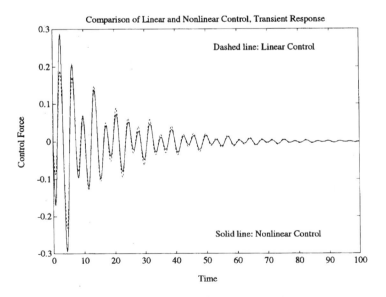

Figure 8. Transient response, control force verses time.

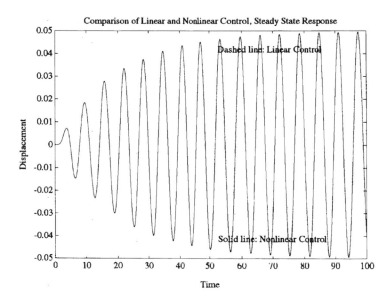

Figure 9. Steady state response, displacement of x_1 verses time. 1 rad/sec. Excitation amplitude = 0.01.

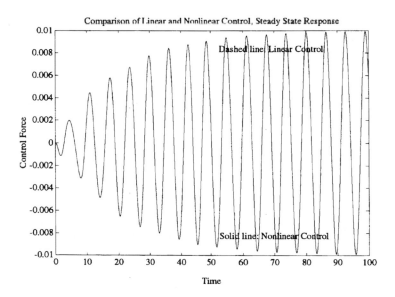

Figure 10. Steady state response, control force verses time 0. 1 rad/sec. Excitation amplitude = 0.01.

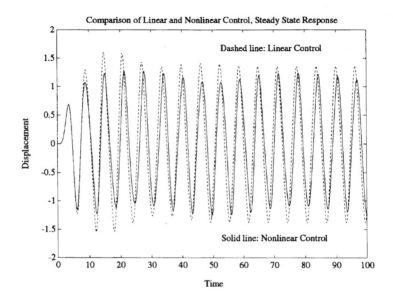

Figure 11. Steady state response, displacement of x_1 versus time 0. 1 rad/sec. Excitation amplitude = 1.

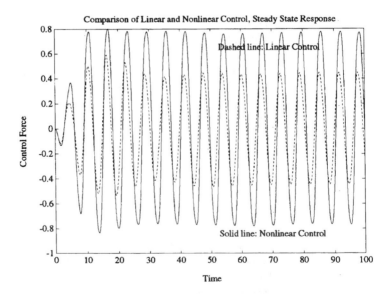

Figure 12. Steady state response; control force versus time. 1 rad/sec. Excitation amplitude = 1.

the consideration of the nonlinear effects in the control design. Although ideal controllability and ideal observability are highly desirable, example 3 demonstrates that for non-ideal controllable systems, the desired control can be approximated and applied. Future work will be in designing nonlinear observers to approximate the states in non-ideally observable systems such that ideal observability is not necessary to fully implement the desired control laws.

Unlike robust nonlinear control methods, such as sliding mode control, the linear modal control approach allows the complete or near complete selection of the nonlinear modal equation dynamics. Although sliding control can be used for vibration suppression, it is intended for use as a tracking control. The purpose of tracking control is to bring the difference between the system outputs and the desired system outputs to zero and maintain the difference at close to zero. Nonlinear modal control is intended as a vibration suppression technique. As such, it allow the explicit modification of the modal equations of motion such that the controlled structure obtains the desired dynamic characteristics something which sliding control is not designed for.

One weakness of nonlinear modal control at this time are that it does not have robust stability with respect to modeling uncertainty in the same way that most linear control methods are not robust. It is thus dependent on obtaining a good model for the structure from which to design the modal control law. Future work will consider improving the robustness of the nonlinear modal control. A second weakness is that, at this time, it is very dependent on near ideal observability. Future work will look into the accuracy of linear observers for the nonlinear system and the use of nonlinear observers in cases where linear observers do not work well. Also note that the modified observability norms are presented because of the absence of nonlinear modal observers at this time (It has not been assumed that if a modal velocity is known that the corresponding modal displacement can be determined since this would require an observer).

One topic for future work include nonlinear modal identification. The dissertation of Vakakis[33] deals with the identification of the modal frequencies and damping of structures with small nonlinearities. In such structures, the deviation between the linear modes and the nonlinear modes is negligible. For structures with larger nonlinearities, the deviation between the nonlinear modes and the linear modes will become greater depending on the nonlinearities (i.e. whether or not the nonlinear modes or similar or non-similar). A possible method for identifying structures with large nonlinearities might be the combination of a nonlinear identification technique, such as minimum model estimation (MME), the nonlinear modal transformations developed by Shaw and Pierre25 and extended here, and some of the concepts of linear modal identification.

Another topic for future work is the application of control via modal structure assignment techniques (The analogous linear methodology is referred to as eigenstructure assignment, but since the nonlinear modes do not correspond to eigenproblems, the term used here is nonlinear modal structure assignment.) The development of the coordinate transformation relations here should allow the modification of not only the modal equations but the mode shapes as well.

Finally, it may be possible to perform the coordinate transformations numerically if the mode shapes can be determined numerically instead of algebraically. This would require that the displacements and velocities of each point on the structure when the structure is moving in a single mode to be written as a second order tensor, hereafter called amplitude tensors. The second order tensor would be spatially discretized with respect to the modal amplitude and modal velocity such that it becomes in effect a look-up table where, for instance, the columns might represent the amplitude dependence with respect to the modal displacement and the rows might represent the amplitude dependence on the modal velocity. For a system with equal phase modes, this would reduce to a first order tensor and for a linear equal phase system, the vector would be equal to the modal amplitude multiplied by element of the mode shape vector corresponding to the given coordinate. Once these tensors have been developed, they could then be effectively substituted for the power series expansions. All matrix inversions and coordinate transformations would then be found numerically using matrices evaluated at each amplitude. Only in the final step would a power series approximation then be used to represent the modal equations of motion, thus eliminating the necessary series truncations and approximations. The results should then be accurate to the amplitude for which the amplitude tensors are found.

References

1. Boivin, N., Pierre, C. and Shaw, S.W., "Non-Linear Normal Modes, Invariance and Modal Dynamics Approximations of Non-Linear Systems," Nonlinear Dynamics, in press.
2. Atkinson, C.P. and Taskett, B., 1965, "A Study of the Nonlinearity Related Solutions of Coupled Nonlinear Systems by Superposition Techniques," *Journal of Applied Mechanics*, **32**, 359–364.
3. Bellman, R., 1960, *Introduction to Matrix Analysis*, McGraw-Hill, New York.
4. Blaquière, A., 1966, *Nonlinear System Analysis*, Academic Press.
5. Caughey, T.K. and O'Kelly, M.E.J., 1965, "Classical Normal Modes in Damped Linear Dynamic Systems," *Journal of Applied Mechanics*, **32**, 583–588.
6. Chen, C.T., 1984, *Linear System Theory and Design*, Holt, Rinehart and Winston, New York.
7. Garcia, E., 1989, *On the Modeling and Control of Slewing Flexible Structures*, Dissertation thesis, University at Buffalo.

8. Greenburg, H.J. and Yang, T.L., 1971, "Modal Subspaces and Normal Mode Vibrations," *International Journal of Nonlinear Mechanics*, **6**, 311–326.

9. Hamdan, A.M.A. and Nayfeh, A.H., 1989, "Measures of Modal Controllability and Observability for First- and Second-Order Linear Systems," *AIAA Journal of Guidance, Control and Dynamics*, **12**(3), 421–429.

10. Hughes, P.C. and Skelton, R.E., 1980, "Controllability and Observability of Linear Matrix-Second-Order Systems," *Journal of Applied Mechanics*, **47**.

11. Inman, D.J., 1989, "Vibration Analysis of Viscoelastic Beams by Separation of Variables and Model Analysis," *Mechanics Research Communications*, **16**, 213–218.

12. Inman, D.J., 1994, *Engineering Vibration*, Prentice Hall, New York.

13. King, M.E. and Vakakis, A.F., "An Energy-Based Formulation for Computing Nonlinear Normal Modes in Undamped Continuous Systems," *Journal of Vibration and Acoustics*, in press.

14. Kryloff, N. and Bogoliuboff, N., 1943, Introduction to Nonlinear Mechanics, Princeton University Press.

15. Minorsky, N., 1969, *Theory of Nonlinear Control Systems*, McGraw Hill.

16. Nayfeh, A.H., Mook, D.T. and Marshall, L.R., 1973, "Nonlinear Coupling of Pitch and Roll Modes in Ship Motions," *Journal of Hydronautics*, **7**, 145–152.

17. Nayfeh, A.H. and Mook, D.T., 1979, *Nonlinear Oscillations*, John Wiley & Sons.

18. Nayfeh, A.H., Nayfeh, S.A., 1994, "On Nonlinear Normal Modes of Continuous Systems," *Journal of Vibration and Acoustics*, **116**, 129–136.

19. Palm, W.J., 1983, *Modeling, Analysis and Control of Dynamic Systems*, John Wiley & Sons, New York.

20. Rand, R.H., 1971, "Nonlinear normal modes in two degree-of-freedom systems," *Journal of Applied Mechanics*, **3B**.

21. Rand, R.H., 1971, "A Higher Order Approximation For Non-Linear Normal Modes in Two Degree of Freedom Systems," *International Journal of Nonlinear Mechanics*, **6**, 545 547.

22. Rand, R.H., 1974, "A Direct Method for Nonlinear Normal Modes," *International Journal of Nonlinear Mechanics*, **9**, 363–368.

23. Rosenberg, R., 1962, "The Normal Modes of Nonlinear n-DOF Systems," *Journal of Applied Mechanics*, **30**, 37–47.

24. Rosenberg, R., 1966, *Advances in Applied Mechanics*, Academic Press, **9**, 155–242.

25. Shaw, S.W. and Pierre, C., 1993, "Normal Modes for Nonlinear Vibratory Systems," *Journal of Sound and Vibration*, **164**, 85–124.

26. Shaw, S.W. and Pierre, C., 1994, "Normal Modes of Vibration for Non-Linear Continuous Systems," *Journal of Sound and Vibration*, **169**, 319–347.

27. Shaw, S.W., 1994, "An Invariant Manifold Approach to Nonlinear Normal Modes of Oscillation," *Journal of Nonlinear Science*.

28. Slater, J.C., 1993, *Nonlinear Modal Control*, Ph.D. Thesis, State University of New York at Buffalo NY.

29. Slater, J.C. and Inman, D.J., 1995, "A Nonlinear Modal Control Method," *Journal of Guidance, Control and Dynamics*, **18**(3), 433–440.

30. Slater, J.C., 1996, "A Numerical Method for Determining Nonlinear Normal Modes," *Journal of Nonlinear Control*.

31. Slotine, J.J.E. and Li, W., 1991, *Applied Nonlinear Control*, Prentice Hall, New York.

32. Szempliska, W., 1990, *The Behavior of Nonlinear Vibrating Systems*, Kluwer Academic Publishers, Boston.

33. Vakakis, A.F., 1990, *Analysis and Identification of Linear and Nonlinear Normal Modes in Vibrating Systems*, Master's thesis, California Institute of Technology.

34. Vakakis, A.F., 1990, *Analysis and Identification of Linear and Nonlinear Normal Modes in Vibrating Systems*, Ph.D. Thesis, California Institute of Technology, Pasadena, California.

35. Vakakis, A.F. and Rand, R.H., 1992, "Normal Modes and Global Dynamics of a Two-Degree-Of-Freedom Non-Linear System-Low Energies," *International Journal of Non-Linear Mechanics*, **27** 861–874.

36. Vakakis, A.F. and Rand, R.H., 1992, "Normal Modes and Global Dynamics of a Two-Degree-Of-Freedom Non-Linear System-High Energies," *International Journal of Non-Linear Mechanics*, **27**, 875–888.
37. Vakakis, A.F., 1992, "Non-Similar Normal Oscillations In a Strong Non-Linear Discrete System," *Journal of Sound and Vibration*, **158**, 341–361.
38. Vakakis, A.F. and Cetinkaya, C., 1993, "Mode Localization in a Class of Multidegree-of-Freedom Nonlinear Systems with Cyclic Symmetry," *SIAM Journal of Applied Mathematics*, **53**, 265–282.
39. Wolfram, S., 1991, *Mathematica, a System for Doing Mathematics by Computer*, Addison-Wesley, New York.
40. Yen, D., 1974, "On the Normal Modes of Nonlinear Dual Mass Systems," *International Journal of Nonlinear Mechanics*, **9**, 45–53.

APPENDIX

The following is the code used to solve the last problem in Mathematica® 2.2. The results must be obtained by executing the code. A copy of this code can be obtained at ftp://norma.cs.wright.edu/pub/nlmcode/nlmcont.ma in ASCII format.

Equations of Motion

VectorDerivative[funs_List,vars_List]:=Outer[D,funs,vars]

The example was originally solved by Shaw and Pierre with two free variables k,g

k=1;g=0.5;

w={{u1},{v1},{u2},{v2}};

z={{x1},{y1},{x2},{y2}};

A[v_]:={{0,1,0,0},{-2-v[[1,1]]^2/2,0,1,0},{0,0,0,1},{1,0,-2,0}}

At[v_]:=A[v].v

The original equations of the system are:

z'==At[z]//MatrixForm

Nonlinear Modal Analysis

The transformation is defined from the nonlinear normal modes as:

M[v_]:={{1,0,1,0},{0,1,0,1},
{1+g((k-3)v[[1,1]]^2-3v[[2,1]]^2)/(2k(k-4)),0,

-1+g((3+7k)v[[3,1]]^2+3v[[4,1]]^2)/(2k(4+9k)),0},
{0,1+3g((k-1)v[[1,1]]^2-v[[2,1]]^2)/(2k(k-4)),0,
-1+3g((1+3k)v[[3,1]]^2+v[[4,1]]^2)/(2k(4+9k))}}

M[w]

where Mode 1 and its dynamics are:

ModeShape1=M[w][[Range[1,4],Range[1,2]]].w[[Range[1,2]]]

Simplify[At[ModeShape1]][[Range[1,2]]]

and Mode 2 and its dynamics are:

ModeShape2=M[w][[Range[1,4],Range[3,4]]].w[[Range[3,4]]]

Simplify[At[ModeShape2]][[Range[1,2]]]

Next we define the transformation matrix and its inverse to find the modal equations of motion

Mt[v_]:=Simplify[M[v].v]

Mteval=Mt[w]

dMtdw=VectorDerivative[Part[Transpose[Mteval],1],
Part[Transpose[w],1]]

At[Mteval]//MatrixForm

Performing the transformation and using e as a small parameter one finds the modal equations of motion. Note if U1 and U2 are of similar magnitude, the dynamics are coupled!! If, however, either U1 or U2 is smaller than the other, the equations reduce to the nonlinear normal mode equations.

EqnMotion=Simplify[ExpandAll[Inverse[dMtdw].At[Mteval]]];

EqnMotion=Simplify[EqnMotion/.{u1->e U1,u2->e U2,v1->e V1,v2->e V2}];

EqnMotion1=Simplify[EqnMotion + O[e]^5]

Physical to Modal Control Transformation

ControlP={{0},{0},{0},{uz}}

{{0}, {0}, {0}, {uz}}

ControlM=Simplify[Inverse[dMtdw].ControlP];

ControlM=Simplify[ControlM/.{u1->e U1,u2->e U2,v1->e V1,v2->e V2}];

ControlM=Chop[ExpandAll[ControlM + O[e]^5]]

r1=Simplify[ControlM[[1,1]]/ControlM[[2,1]] + O[e]^5]

r2=Simplify[ControlM[[3,1]]/ControlM[[4,1]] + O[e]^5]

M0={{1,0,1,0},{0,1,0,1},{1,0,-1,0},{0,1,0,-1}}

Control

M2[v_]:=M[v]-M0;

IM0=Inverse[M0]; I4=IdentityMatrix[4];

Nn[v_]:=(I4-IM0 . M2[IM0 . v]).IM0

Nt[v_]:=Simplify[Nn[v].v]

Simplify[Nt[Mt[w]]] /. {u1->e U1,u2->e U2,v1->e V1,v2->e V2};
Simplify[ExpandAll[%] + O[e]^5]

Nteval=Nt[z];
dNtdz=VectorDerivative[Part[Transpose[Nteval],1],
Part[Transpose[z],1]];

func1[v_]:=-.1v[[2,1]](1+v[[2,1]]^2+v[[1,1]]^2)+
0.1v[[4,1]](1+v[[4,1]]^2+3 v[[3,1]]^2)

func2[v_]:=-0.25 v[[1,1]]v[[2,1]]+0.0576923v[[3,1]]v[[4,1]]

Control[v_]:={{func1[v]func2[v]},{func1[v]},
{-func1[v]func2[v]},{-func1[v]}}

CNteval=Simplify[Control[Nteval]];

IdNtdz=Simplify[Inverse[dNtdz]];

ControlP=IdNtdz.CNteval;

ControlP=ControlP/.{x1->e X1,x2->e X2,y1->e Y1,y2->e Y2};

ControlP=Chop[ExpandAll[ControlP + O[e]^5]]

4 SUPERELEMENT MODELING OF VEHICLE DYNAMIC STRUCTURAL SYSTEMS

O.P. AGRAWAL[1,*], K.J. DANHOF[2] and R. KUMAR[3]

[1]*Department of Mechanical Engineering, Southern Illinois University, Carbondale, IL 62901, USA*
[2]*Department of Computer Science, Southern Illinois University, Carbondale, IL 62901, USA*
[3]*Sun Microsystems, Inc., 2550 Garcia Ave., MTV19-215, Mountain View, CA 94043-1109, USA*

This paper presents a superelement model based parallel algorithm for a planar vehicle dynamics. The vehicle model is made up of a chassis and two suspension systems of which each consists of an axel-wheel assembly and two trailing arms. In this model, the chassis is treated as a Cartesian element and each suspension system is treated as a superelement. The parameters associated with the superelements are computed using an inverse dynamics technique. Suspension shock absorbers and the tires are modeled by nonlinear springs and dampers. Euler-Lagrange approach is used to develop the system equations of motion. This leads to a system of differential and algebraic equations in which the constraints internal to superelements appear only explicitly. The above formulation is implemented on a multiprocessor machine. The numerical flow chart is divided into modules and the computation of several modules is performed in parallel to gain computational efficiency. In this implementation, the master (parent processor) creates a "pool" of slaves (child processors) at the beginning of the program. The slaves remain in the pool until they are needed to perform certain tasks. Upon completion of a particular task, a slave returns to the pool. This improves the overall response time of the algorithm. The formulation presented is general which makes it attractive for a general purpose code development. Speedups obtained in the different modules of the computer code are also presented. Results show that the superelement model based parallel algorithm can significantly reduce the vehicle dynamics simulation time.

*Correspondence.

169

4.1. INTRODUCTION

Dynamic simulation and control of ground vehicles require access to high speed computational engines and efficient analytical formulations and numerical algorithms, especially when real-time simulation, modeling of flexibility and crashworthiness, man-in-the-loop, and instant graphic visualizations are considered. Significant progress has been made in both areas formulations and algorithms, and computational hardware tools. Current formulations are several times faster than those developed 20 years ago due to extensive progress made in the area of multibody dynamics. Recent emergence of high-speed computational tools and environments such as vector, multi- and massively parallel processors, multi-threading operating systems, computer networking, and high speed graphics engines have considerably changed the way vehicle systems are analyzed. Vector, multi- and massively parallel processor machines allow execution of several instructions in parallel. Networking allows computation of different modules associated with a simulation in parallel on different remote machines. Usage of concurrent programming paradigms based on user level threads, for example, as implemented in SunOS/Solaris operating system, do help exploit the newer and faster multiprocessor machines to a varying degree depending on how much the uniprocessor software base has been modified. High speed graphics engines allow an instant display of simulation results. All this has increased the computational speed. These advances in computer software and hardware technologies have motivated investigators to develop new vector-parallel algorithms for vehicle dynamics, and the real-time simulation and control including real-time graphical visualization and man-in-the-loop are now a reality.

Although progress has been made in the area of parallel algorithms for real-time simulation of vehicle systems, much remains to be done. Development of efficient parallel algorithms for the above tasks relies on two major efforts: the efficient utilization of current software, and hardware technologies and the efficient analytical formulations for the systems. The distribution of computational load among all processors should be such that the processor-idle-time is minimum, and the analytical formulation should be such that it leads to computation of minimum number of terms.

This paper presents a superelement based parallel algorithm that incorporates both efforts in the dynamic simulation of a planar vehicle. The basic ideas introduced here are: 1) represent a group of elements forming a single-degree-of-freedom subsystem by one element called a superelement, and represent the properties of this element by a set of single parameter (called a superelement parameter) curves. 2) identify the tasks that can be performed in parallel, and develop an algorithm that distributes the tasks among processors in an efficient manner. The superelement parameter curves eliminate the need for numerous repetitive and unnecessary calculations. The parallel algorithm implemented here is based on master slave model in which the master processor creates a "pool" of slaves at the beginning of the program. The slaves remain in the pool until they are needed to perform certain tasks. When tasks arrive, then, depending on their IDs, the slaves are released from the pool. It is shown that the algorithm leads to improvement in the simulation response time. The formulation is general and it may be used in dynamic simulations of a large class of multibody systems.

4.2. BACKGROUND

This section presents a brief review of research in the area of vehicle dynamics, multibody dynamics, and the role of parallel computers and algorithms in these areas. Several investigators have presented formulations and simulation results for vehicle dynamics and control.[1-8] References[9 10] present reviews of computer codes for vehicle dynamics. In the past, inertia components of a vehicle were treated as rigid. However, in recent years, several investigators have presented flexible models for various vehicles.[11-15] This allows simulation of a crash and/or an overturn and determination of major flaws/errors in the design of a vehicle at a very low cost without injury or loss of life or instruments. Crash and overturn simulation results for vehicles appear in[16-21]. Sayers and coworkers[22-24] have presented symbolic codes for simulating complex automobiles. The advantages of symbolic codes are that 1) they eliminate/reduce errors that may result due to human oversight, and 2) the resulting codes can be tailored to a specific vehicle thus improving the computational efficiency. The above research is not limited to road vehicles. References[25-31] provide formulations and results for railway vehicles. Formulations and simulation results for tracked vehicles and soil-track interaction appear in[32,33]. It is realized that software development is a dynamic process and the sequential codes are not suitable, especially when reusability and expandability are of major concern. Although procedural based codes attempt to accomplish the above goal, a number of current computer software experts believe that a new approach to programming called Object Oriented Programming (OOP) is very efficient when reusability and expandability issues are important. Daberkow and Schiehlen[34] have used OOP concepts to develop DAMOS-C for multibody systems including vehicle dynamics.

The above list of references describing formulations, algorithms, and simulation results for vehicle dynamics is by no means complete and there are many more references available in this area. For dynamic simulation purposes a vehicle is treated as a multibody system. As a matter of fact, numerous MB codes are routinely being used for vehicle dynamics. Therefore, a review of multibody literature can help improve the way the vehicle simulations are performed.

Considerable work has been done in the past twenty years in the area of computer-aided analysis of multibody systems.[35-52] A general survey indicates that during this period over 1000 papers were published in this area. A general history of the work done in this area prior to 1977 is given in[53]. Other surveys include[54], mainly on machine dynamics;[55], largely on formulation of Lagrangian type;[56], on multi-body spacecraft dynamics;[53], on Eulerian formulation; and[57], on Lagrangian and Eulerian methods for systems with tree structures, but not on their computer implementations. Various methods are compared in[58] without reference to computational aspects. More general surveys are found in[59], mainly on rigid multi-body dynamics, and in[60] on mixed rigid and flexible multi-body system dynamics.

The field of multibody dynamics has progressed to a point where general purpose commercial multibody dynamics codes are now available. Several codes such as ADAMS, COMPAMM, DADS, DAMS, MESA-VERDE, and MOBILE are now routinely being used in industries and academia for the analysis and design of multibody systems. In the introduction of reference[61], Schiehlen provides a detail comparison of 20 codes for multibody systems. The comparison includes the topology considered; the mechanical elements supported; the coordinates used in the formulation; the computations involved such as numeric or symbolic or both; the approach taken to obtain the differential equations; the type of resulting equations such as linear or nonlinear or mixed, stiff or nonstiff, and differential or differential-algebraic; pre- and post-processing capabilities available; the language used to develop the codes, the operating systems and the computers for which the codes were developed; and various applications supported by the code. Reference[61] also includes a set of papers by the original authors describing theory and formulations of their codes and results for some simulations. Reference[62] provides a chronological survey of some general purpose computer programs for the analysis of mechanical systems. Several authors have written books dedicated to this area. They are Wittenburg[63], Nikravesh[64] and Haug[65], mainly on rigid multibody systems, and Shabana[66], Huston[67], and Amirouche[68], on mixed rigid-flexible multibody systems. Reference[67] also provides an extensive list of references in this area.

Recently, several investigators have used recursive approaches for the analysis of multibody systems. Li[69] has shown that the dynamic response of a six-degree-of-freedom manipulator, which took over 117,000 mathematical operations 27 years ago, can now be obtained in less than 1,500 operations.

Reference[69] also provides a comparison of several dynamics formulations. Other investigations incorporating recursive approaches include[70-72], on rigid open-loop;[73-77], on rigid close-loop; and[78], on rigid/flexible multibody systems. A comprehensive treatment that employs minimum generalized coordinates for dynamics of multibody systems appears in[79,80].

The size and capability of multibody codes have grown, so has the problem of managing these codes. Current multibody codes not only perform dynamic simulations in stand alone mode, but they also communicate (or they will be expected to communicate) with other computer aided analysis and design codes. Object Oriented Programming Approach (OOPA) is now being used to overcome this and similar software maintenance issues. References[34,81] among others provide OOPA for multibody dynamics.

References[82-84] develop the idea of superelements for multibody dynamics. The superelement defined here is a group of elements forming a single degree-of-freedom with respect to its reference frame. The concept of superelement is important since the inertia and other properties of a superelement may be represented as a set of one parameter curves. These properties do not depend on the dynamics of the system. Therefore, these curves (or properties) may be generated well in advance or as the simulation proceeds. Once these properties are known, a superelement may be treated just like an object. Reducing a number of elements into one element significantly reduces the dimension of the problem and the CPU time.[82-84] Note that a superelement in object oriented modeling approach can be treated like any other element. In this respect, the concept of superelement defined here is very similar to single degree of freedom kinematic transfer elements defined in[79-81].

Computer response time can be improved considerably with the help of vector and parallel machines and new advance features such as multi-threading in modern operating systems . Gluck[85] has indicated that on some machines the ratio of sequential to parallel simulation time for flexible spacecraft systems can easily exceed a factor of 800. In contrast to analytical formulations, very few papers have appeared in the area of multibody dynamics that take full advantage of these new high speed engines. Dubetz et al.[86] present a real-time dynamic simulation scheme that takes full advantage of computing resources over a network. Tak and Kim[87] present a parallel processing scheme for design sensitivity analysis of multibody systems. Tsai and Haug[88] and Bae et al.[89] have presented parallel algorithms for multibody dynamic simulations. Chang and Kim[90] have used a recursive parallel scheme for real-time simulation of multibody systems with man in the loop. Kurdila and Kamat[91] have presented a concurrent multiprocessing algorithm for calculating orthonormal bases for simulating the dynamics of multibody systems.

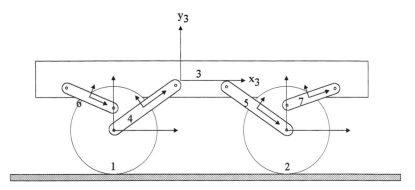

Figure 1. Planar Vehicle model.

Although the recursive parallel and concurrent multiprocessing orthonormal bases algorithms presented above improve the computational speeds, they do not take full advantage of superelements present in the systems. As a result, in these techniques the superelement properties are computed implicitly many times which could have been avoided using the parameterized curves.

From the above discussion, it is clear that for vehicle dynamics a superelement model based parallel algorithm will provide some additional advantages over current parallel methods. In this paper, one such algorithm for a planar vehicle is presented. Mathematical derivations presented are largely tailored to this vehicle. However, the approach presented is general and can be extended to arbitrary vehicle systems.

4.3. THE PLANAR VEHICLE MODEL

The kinematic configuration of the planar vehicle model considered here is the same as that of Ref.[83]. For completeness and ease in the discussion to follow, this model is briefly described here.

The planar vehicle model (Figure 1) consists of two front and rear axle-wheel assemblies (bodies 1 and 2), one chassis (body 3), and four trailing arms (bodies 4 to 7) that connect the chassis and wheel assemblies by revolute joints. The function of the trailing arms is to provide kinematic control of the axle position and to absorb driving and braking torques acting on the wheels. Therefore, they are modeled as a Watt's mechanism, which gives very small rotation to the axle during vertical displacement. Each axle-wheel mass is assumed to be concentrated at the wheel center. The mass of the chassis

Table 1. Inertia properties of the planar vehicle model

Body no.	Mass (kg)	Moment of inertia about C.G. (kg-m²)
1	1185	13.33
2	1185	13.33
3	11950	58300
4	22.4	2.38
5	22.4	2.38
6	17.6	1.14
7	17.6	1.14

includes masses of the payload and the engine. Mass and moment of inertia of each body are given in Table 1.

Suspension springs, dampers and the tires are modeled by springs and dampers as shown in Figure 2. Spring characteristics of tires are taken as quadratic functions of displacement ($K_t = 11.298E07$ N/m², $C_t = 9250$ Ns/m, Tire radius = 0.6 m). A simple point contact tire model is used to simulate tire forces that occur due to motion of wheel relative to the road surface. The ground-tire force is assumed to be vertical and the fore and aft force components are neglected. The tire is free to leave the ground, to simulate the wheel hop. Bilinear spring and damping characteristics of suspension elements considered are as follows: spring rate $K_s = 1.382 * 10^6$ N/m, and the damping coefficients $C_{sc} = 10960$ Ns/m in compression and $C_{sr} = 35150$ Ns/m in rebound. These characteristics are shown in Figure 3. The high stiffness of suspension spring in compression, when the spring deflection is greater than 0.15m, simulates the bump-stop in the suspension system.

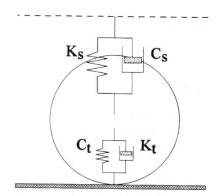

Figure 2. Suspension and tire model.

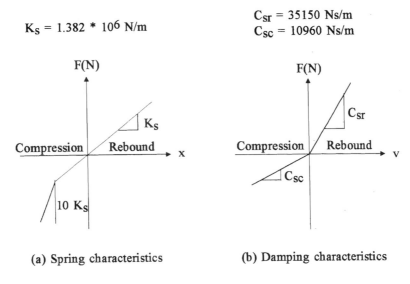

(a) Spring characteristics   (b) Damping characteristics

Figure 3. Suspension characteristics.

Note that both suspension systems have one degree of freedom with respect to the chassis and therefore they are treated as superelements.

4.4. CARTESIAN ELEMENT MODELING

One of the key steps in modeling the dynamics of a vehicle and in general the dynamics of a general multibody system is to define the configuration of an element of the system in a given coordinate frame. Once this is done, expressions for constraints and differential equations, describing kinematic compatibility between adjacent elements (or the topology) and the dynamics, respectively, of the system, can be developed in a systematic way. For a comprehensive treatment of the subject, a Cartesian coordinate scheme to describe the configuration of a rigid link is considered first. When the configurations of all links in the system are described using Cartesian coordinate only, the formulation is called a Cartesian element modeling.

To acomplish the above task, consider a body i in the global X-Y Cartesian coordinate system, as shown in Figure 4. Here $\xi_i - \eta_i$ is the local body coordinate system rigidly attached to body i. One of the advantages of using body coordinate systems is that in this system the locations of all points in the body remain fix. Let o_i be the origin of the body frame i. In the discussion to

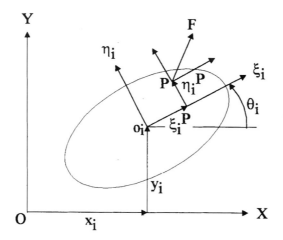

Figure 4. Generalized coordinates of body i.

follow, it is assumed that o_i is attached to the center of mass of body i because this leads to several simplifications. The case where o_i is not attached to the center of mass of body i is left as an exercise for the reader.

Using Cartesian coordinates, the configuration of body i (also called a Cartesian element) may be defined with respect to the global reference frame as

$$s_i = [x_i, \ y_i, \ \theta_i]^T \tag{1}$$

where x_i and y_i are the global Cartesian coordinates of point o_i, and θ_i is the orientation of the axis $o_i\xi_i$ with respect to the global X-axis. Once s_i is defined, the location of Point P (Figure 4) is given as

$$
\begin{aligned}
x^P &= x_i + \xi_i^P \cos(\theta_i) - \eta_i^P \sin(\theta_i) \\
y^P &= y_i + \xi_i^P \sin(\theta_i) + \eta_i^P \cos(\theta_i)
\end{aligned}
\tag{2}
$$

where ξ_i^P and η_i^P are the coordinates of Point P in Body Frame i. Assuming that the global Y-axis is vertical, expressions for the kinetic and potential energy terms for the element can be written, respectively, as

$$T = \frac{1}{2}m_i(\dot{x}_i^2 + \dot{y}_i^2) + \frac{1}{2}I_i\dot{\theta}_i^2 \tag{3}$$

and

$$V_i = m_i g y_i \tag{4}$$

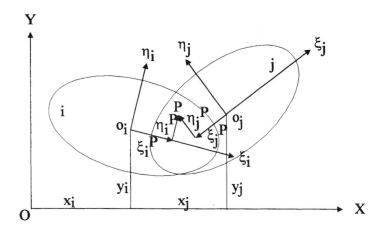

Figure 5. Revolute joint constraints.

where m_i and I_i are the mass and the moment of inertia of body i, g is the gravitational constant, and the period on (*) represents time derivative of (*).

Let F be a force vector acting at point P as shown in Figure 4. The generalized force components corresponding to this force are

$$F_x = F_{x_i^P} \cos(\theta_i) - F_{y_i^P} \sin(\theta_i)$$
$$F_y = F_{x_i^P} \sin(\theta_i) + F_{y_i^P} \cos(\theta_i) \tag{5}$$
$$F_\theta = -F_{x_i^P} \eta_i^P + F_{y_i^P} \xi_i^P$$

where $F_{x_i^P}$ and $F_{y_i^P}$ are the components of F along the ξ_i and η_i axes, F_x and F_y are the components of F along the X and Y axes, and F_θ represents the torque about the center of mass.

The revolute joint constraints (or conditions) between two adjacent bodies i and j (Figure 5) are written as

$$(x_i + \xi_i^P \cos(\theta_i) - \eta_i^P \sin(\theta_i)) - (x_j + \xi_j^P \cos(\theta_j) - \eta_j^P \sin(\theta_j)) = 0$$
$$(y_i + \xi_i^P \sin(\theta_i) + \eta_i^P \cos(\theta_i)) - (y_j + \xi_j^P \sin(\theta_j) + \eta_j^P \cos(\theta_j)) = 0 \tag{6}$$

Constraint conditions for other type of joints can be developed in a similar manner.[64-66] Once the above expressions are defined, the equation of motion of the entire vehicle system can be written using the Lagrange multiplier approach as

$$\frac{d}{dt} \left(\frac{\partial T}{\partial \dot{q}} \right)^T + \left(\frac{\partial \phi}{\partial q} \right)^T \lambda = Q - \left(\frac{\partial V}{\partial q} \right)^T \tag{7}$$

where $q = [x_1,\ y_1,\ \theta_1\ \cdots]^T$ is the vector of generalized coordinates, $T = \sum_i T_i$ and $V = \sum_i V_i$ represent the total kinetic and the total potential energy terms, ϕ, λ and Q are the vectors of constraints, Lagrange multipliers, and generalized forces, respectively, and $\partial\phi/\partial q$ represents the constraint Jacobian matrix.

Substituting Equations (3) and (4) into Equation (7), the equations of motion can be written as

$$M\ddot{q} + \left(\frac{\partial\phi}{\partial q}\right)^T \lambda = Q + Q_V \tag{8}$$

where M is a diagonal system mass matrix and Q_V is the vector of gravitational force which is given as

$$Q_V = [0\ \ m_1 g\ \ 0\ \ 0\ \ m_2 g\ \ \cdots]^T. \tag{9}$$

Equation (9) can also be obtained using other methods such as a virtual work approach.[66]

Observe that the vehicle considered here has 7 bodies and 8 revolute joints, and therefore, in Cartesian modeling, q and ϕ will be of dimensions 21×1 and 16×1. Connectivity between adjacent links to generate the constraints equations for the vehicle system is given in Table 2. Constraint equations play a significant role in position, velocity, and acceleration analyses of the system. The resulting differential and constraint equations can be solved using various numerical methods.[36,39,40,44,52]

4.5. LAGRANGIAN ELEMENT MODELING

When the configuration of a body in a planar motion is described by less than 3 generalized coordinates associated with the body and perhaps generalized coordinates of other bodies, these coordinates are called Lagrangian coordinates and the body is called a Lagrangian element. Observe that there is no joint internal to this element.

Several possibilities exist for Lagrangian coordinates. For example, the Lagrangian coordinates of an element may be a subset of Cartesian coordinates, a combination of Cartesian coordinates, or a set of coordinates describing the relative degrees of freedom of the element with respect to its adjacent element. The last set has been selected by many investigators because it leads to efficient recursive formulation. Since an element in a system may be connected to several other elements, this set is not well defined. To overcome this difficulty, the system is transformed into a tree by removing minimum number of joints. In the resulting tree, each body has

Table 2. Connectivity between adjacent links

Joint no.	1	2	3	4	5	6	7	8
Link i	1	1	2	2	3	3	3	3
Link j	4	6	5	7	4	6	5	7

only one predecessor body. Once this is done, the Lagrangian coordinates of an element are defined with respect to its predecessor body. Thus the number of Lagrangian coordinates of an element also give the relative degrees of freedom of the element with respect to its predecessor element. Compatibility conditions at the cut joints give the necessary constraint equations.

Observe that the vehicle system considered here will lead to a total of 9 Lagrangian coordinates and 4 constraint equations. This is because in this formulation some of the constraints are eliminated explicitly. For further details on a Lagrangian element formulation for mechanical systems, readers are refered to[84].

4.6. SUPERELEMENT MODELING

A superelement is defined here as a group of constrained elements that forms a single degree-of-freedom with respect to its reference frame. In this respect, a superelement may be a part of a mechanical system or an entire mechanical system. In the reference frame, the configuration of the superelement is defined using a parameter ψ (called a superelement parameter or a superelement coordinates). This parameter may be one of the Cartesian coordinates[36], the Lagrangian coordinates[84], or a combination of coordinates, such as, the tangent coordinates.[39,44,45,52]

In order to develop the equation of motion of a superelement in a moving reference frame, consider a typical link i of the superelement and the reference frame x_r-y_r associated with the superelement as shown in Figure 6. Point c in Figure 6 is the center of mass of link i. Using the approach presented in section 4, the configuration of link i in its reference frame is defined as

$$s_i = [\, x_i(\psi), \quad y_i(\psi), \quad \theta_i(\psi) \,]^T \tag{10}$$

where x_i and y_i are the Cartesian coordinates of point C and θ_i is the orientation of the axis $C\xi_i$ with respect to the reference frame. Note that x_i, y_i, and θ_i have implicitly been represented as functions of ψ. This is possible mainly because the superelement has only one degree-of-freedom with respect to its reference frame. The configuration of link i may alternatively be defined and a corresponding formulation may be developed using Lagrangian coordinates. For simplicity, Cartesian coordinates are used here to represent internal elements of a superelement.

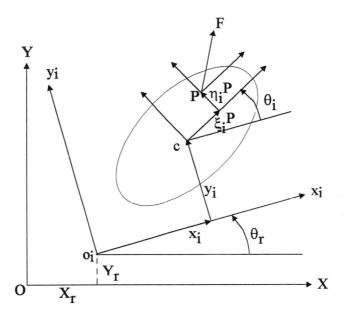

Figure 6. Generalized coordinates of body i in a moving frame.

Using Figure 6, the global coordinates of point P may be written as

$$r_i = \begin{bmatrix} X_r \\ Y_r \end{bmatrix} + A(\theta_r) \begin{bmatrix} x_i \\ y_i \end{bmatrix} \tag{11}$$

where X_r, Y_r are the global x, y locations of the origin or the moving reference frame, and $A(\theta_r)$ is the transformation matrix defined as

$$A(\theta_r) = \begin{bmatrix} \cos(\theta_r) & -\sin(\theta_r) \\ \sin(\theta_r) & \cos(\theta_r) \end{bmatrix} \tag{12}$$

Here θ_r is the angle of rotation of the reference frame with respect to the global frame. Using Equation (11), the velocity of point P can be written as

$$v_i = \dot{r}_i = \begin{bmatrix} \dot{X}_r \\ \dot{Y}_r \end{bmatrix} + A(\theta_r) \begin{bmatrix} \partial x_i / \partial \psi \\ \partial y_i / \partial \psi \end{bmatrix} \dot{\psi} + \dot{A}(\theta_r) \begin{bmatrix} x_i \\ y_i \end{bmatrix} \tag{13}$$

Using Figure 6, the angular velocity of link i ($= \omega_i$) with respect to the global frame may be written as

$$\omega_i = \dot{\theta}_r + \dot{\theta}_i \tag{14}$$

Let the superelement contain n_l number of such links. The kinetic energy (T) for the superelement is obtained by summing the kinetic energies of each of its elements. Thus,

$$T = \sum_{i=1}^{n_l} \left[\frac{1}{2} m_i \dot{r}_i^T \dot{r}_i + \frac{1}{2} I_i \dot{\omega}_i \right] \qquad (15)$$

Using Equations (13) and (14), Equation (15) may be written in a more compact form as

$$T = \frac{1}{2} \dot{q} M(\theta_r, \psi) \dot{q} \qquad (16)$$

where

$$q = [X_r \quad Y_r \quad \theta_r \quad \psi]^T \qquad (17)$$

is a vector of generalized coordinates, and $M(\theta_r, \psi)$ is a symmetric 4×4 superelement mass matrix. The elements of mass matrix M are given in[92]. Note that the matrix M depends on the rotational coordinate θ_r and the superelement parameter ψ.

The potential energy (V) for the superelement may be written in a similar fashion, as

$$V(Y_r, \theta_r, \psi) = g \sum_{i=1}^{n_l} m_i \left[Y_r + x_i(\psi)sin\theta_i(\psi) + y_i(\psi)cos\theta_i(\psi) \right] \qquad (18)$$

In this case, the potential energy does not depend on X_r because the global x-axis is horizontal. In general, the potential energy is a function of X_r also. Let

$$Q = [Q_{X_r} \quad Q_{Y_r} \quad Q_{\theta_r} \quad Q_\psi]^T \qquad (19)$$

be a vector of generalized force associated with non-conservative and non-gravitational forces. To demonstrate a method for obtaining the vector Q, consider a force F acting at point P as shown in Figure 6. The location of point P may be written as

$$r_P = \begin{bmatrix} X_r \\ Y_r \end{bmatrix} + A(\theta_r) \begin{bmatrix} x_i \\ y_i \end{bmatrix} + A(\theta_r + \theta_i) \begin{bmatrix} \xi_i^P \\ \eta_i^P \end{bmatrix} \qquad (20)$$

where ξ_i^P and η_i^P are the coordinates of point P in the body fixed frame. The virtual work due to this force may be written as

$$\delta w = \delta r_p^T F = \delta q^T Q \qquad (21)$$

where $\delta(*)$ represents the variation of a parameter. Here, vector Q provides the generalized force associated with force F. The components of Q are given as

$$\begin{bmatrix} Q_{X_r} \\ Q_{Y_r} \end{bmatrix} = F \qquad (22a)$$

$$Q_{\theta_r} = F^T \frac{\partial A(\theta_r)}{\partial \theta_r} \begin{bmatrix} x_i \\ y_i \end{bmatrix} + F^T \frac{\partial A(\theta_r + \theta_i)}{\partial \theta_r} \begin{bmatrix} \xi_i^P \\ \eta_i^P \end{bmatrix} \qquad (22b)$$

$$Q_\psi = F^T A(\theta_r) \begin{bmatrix} \partial x_i/\partial \psi \\ \partial y_i/\partial \psi \end{bmatrix} + F^T \frac{\partial A(\theta_r + \theta_i)}{\partial \theta_i} \begin{bmatrix} \xi_i^P \\ \eta_i^P \end{bmatrix} \frac{\partial \theta_i}{\partial \psi} \qquad (22c)$$

Generalized force associated with other forces and torques may be obtained in a similar fashion.

Consider the case where there is no external kinematic constraints on the superelement and the reference frame is completely free to move in the plane. Using Lagrange's approach, the differential equations of motion of this system are given as[66]

$$\frac{d}{dt} \left(\frac{\partial T}{\partial \dot{q}} \right)^T - \left(\frac{\partial T}{\partial q} \right)^T = Q - \left(\frac{\partial V}{\partial q} \right)^T \qquad (23)$$

Observe that compared to Equation (7), Equation (23) does not contain the constraints internal to the superelement but in contains an additional term $\partial T/\partial q$. This is because these constraints have implicitly been eliminated thus making the kinetic energy dependent on q.

Substituting Equations (16), (18), and (21) into Equation (23) yields the differential equations of a superelement in a moving frame as

$$M\ddot{q} + H + \left(\frac{\partial V}{\partial q} \right)^T = Q \qquad (24)$$

where vector H is defined as

$$H = \left[\frac{\partial(M\dot{q})}{\partial q} - \frac{1}{2} \left(\frac{\partial(M\dot{q})}{\partial q} \right)^T \right] \dot{q} \qquad (25)$$

Note that the differential equations of motion of a superelement consisting of several elements have been written only in terms of coordinates of the reference frame and the superelement parameter.

4.7. DYNAMICS EQUATIONS FOR THE VEHICLE

From Figure 1, it is clear that the two suspension systems have one degree-of-freedom each with respect to the chassis. Therefore, each suspension system can be treated as a superelement. Using the chassis as the reference frame for the superelements (or the suspension systems), the kinetic energy T and the potential energy V of the vehicle system may be written as

$$T = \frac{1}{2} \sum_{k=1}^{2} (\dot{q}_k)^T M_k(q_r, \psi_k) \dot{q}_k + \frac{1}{2} \dot{q}_r^T M_r(q_r) \dot{q}_r \tag{26}$$

and

$$V = \sum_{k=1}^{2} V_k(q_r, \psi_k) + V_r(q_r) \tag{27}$$

where q_k, M_k, V_k, and ψ_k are, respectively, the generalized coordinate vector, the mass matrix, the potential energy, and the superelement coordinate associated with superelement k; and q_r, M_r, and V_r are, respectively, the generalized Cartesian coordinate vector, the mass matrix, and the potential energy associated with the reference frame. The expressions for the kinetic and the potential energy terms have been written implicitly. These expressions must be obtained either analytically or numerically. Note that there is no coupling between the two superelement terms. Therefore, several terms associated with a superelement may be computed in parallel with that of the other.

In order to get a further insight into the formulation, vector q_k and matrix M_k are partitioned as

$$q_k = \begin{bmatrix} q_r \\ \psi_k \end{bmatrix} \tag{28}$$

and

$$M_k = \begin{bmatrix} M_k^{11} & M_k^{12} \\ M_k^{21} & M_k^{22} \end{bmatrix} \tag{29}$$

where submatrices $(M_k^{11})_{3 \times 3}$ and $(M_k^{22})_{1 \times 1}$ are contributions of mass matrix M_k to reference coordinates q_r and superelement coordinate ψ_k, and $(M_k^{12})_{3 \times 1}$ and $(M_k^{21})_{1 \times 3}$ are the coupling mass matrices. These matrices define inertia properties of the element. In[92], it is shown that the submatrix $(M_k^{22})_{1 \times 1}$ is completely free of reference coordinates. Both reference and superelement coordinates appear in submatrices $(M_k^{11})_{3 \times 3}$, $(M_k^{12})_{3 \times 1}$ and $(M_k^{21})_{1 \times 3}$ but only in isolated form. This partitioning makes it possible to identify and compute superelement terms independent of reference terms.

The composite vector of coordinates, called vector of system generalized coordinates may now be defined as

$$q = [\, q_r^T \quad \psi_1 \quad \psi_2 \,]^T \tag{30}$$

Note that the dimension of the vector of generalized coordinates in superelement formulation is smaller than that of relative (or Lagrangian) coordinate formulation[84], and much smaller than the Cartesian element formulation.[36] Let

$$\Phi(q, t) = 0 \tag{31}$$

be a vector of constraints. This vector does not contain any constraint internal to a superelement. In a general multibody systems, such constraints may appear due to many reasons.[83] Although it may seem that such constraints will not appear in the current vehicle system, it is not so. This is because in vehicle dynamics simulation, it is quite common to constrain the horizontal (or some other) motion of the chassis, and move the ground or the terrain underneath the vehicle. Such restrictions provide constraint equations. If constraints other than constraints internal to superelements are not present, Equation (31) may be neglected.

Using the Lagrange multiplier approach, the differential equations of motion of the system are given as[66]

$$M\ddot{q} + H + \left(\frac{\partial V}{\partial q}\right)^T + \left(\frac{\partial \Phi}{\partial q}\right)^T \lambda = Q \tag{32}$$

where

$$M = \begin{bmatrix} M_{(1)}^{11} + M_{(2)}^{11} + M_r & M_{(1)}^{12} & M_{(2)}^{12} \\ M_{(1)}^{21} & M_{(1)}^{22} & 0 \\ M_{(2)}^{21} & 0 & M_{(2)}^{22} \end{bmatrix} \tag{33}$$

is the system mass matrix,

$$H = [\, (H_r^{(1)} + H_r^{(2)} + H_r)^T \quad H_{\psi_1} \quad H_{\psi_2} \,]^T \tag{34}$$

is a vector of quadratic velocity terms,

$$Q = [\, (Q_r^{(1)} + Q_r^{(2)} + Q_r)^T \quad Q_{\psi_1} \quad Q_{\psi_2} \,]^T \tag{35}$$

is a system generalized force vector due to nonconservative forces and conservative forces that are not included in the potential energy functions, and λ is the vector of Lagrange multipliers. Equation (32) provides a set of differential equations of motion for the vehicle system that are much less in number in comparison to the Cartesian and the Lagrangian coordinate formulations.[36,84] Differential equations of motion for a general class of systems are obtained in a similar manner. Note that each superelement in the system is treated as one element. Furthermore, parameters associated with a superelement are independent of parameters associated with other superelements. Therefore, the present scheme is suitable for use with parallel processors to improve the computational efficiency. A parallel algorithm for this is presented in the next section.

Let $t = t_0$ be the initial time and

$$q(t_0) = q_0, \tag{36}$$

and

$$\dot{q}(t_0) = \dot{q}_0, \tag{37}$$

be the initial conditions which are consistent with the constraint Equation (31). Equations (31), (32), (36) and (37) define a set of differential and algebraic equations. These equations may be solved using one of the schemes discussed in[32,39,40,44,52], among others, provided that the superelement parameters are known.

4.8. COMPUTATION OF SUPERELEMENT PARAMETERS

Earlier in this section, it was stated that the superelement terms may be obtained using an analytical or a numerical approach. In many applications it may not be feasible to obtain closed form analytical expressions for these terms and a numerical approach must be considered. A detailed numerical approach for obtaining superelement parameters appear in[92]. A brief account of the approach is presented here.

In order to demonstrate a method for obtaining superelement functions, consider that the chassis and the second superelement are fixed. In this case, Equation (32) leads to the following equation:

$$M_{(2)}^{22} \ddot{\psi}_1 + \frac{1}{2} \frac{d M_{(2)}^{22}}{d \psi_1} \dot{\psi}^2 + \frac{d V_{(1)}}{d \psi_1} = Q_{\psi_1} \tag{38}$$

Here, $M_{(2)}^{22}$, $d M_{(2)}^{22} / d \psi_1$, and $d V_{(1)} / d \psi_1$ are some of the superelement terms which are the same as those in Equation (32). Note that these terms depend only on the superelement parameter ψ_1. The inverse dynamics method for obtaining the above superelement terms is as follows[83]: Consider three cases, first $\psi_1 = \psi_{10}$, and $\dot{\psi}_1 = \ddot{\psi}_1 = 0$, second $\psi_1 = \psi_{10}$, $\dot{\psi}_1 = 1$, and $\ddot{\psi}_1 = 0$, and third $\psi_1 = \psi_{10}$, and $\dot{\psi}_1 = \ddot{\psi}_1 = 1$. Let the three generalized forces corresponding to these three cases be $Q_{\psi_1(1)}$, $Q_{\psi_1(2)}$, and $Q_{\psi_1(3)}$. From Equation (38), it then follows that

$$\frac{dV_{(1)}}{d\psi_1} = Q_{\psi_1(1)} \tag{39}$$

$$\frac{1}{2}\frac{dM_{(2)}^{22}}{d\psi_1} = Q_{\psi_1(2)} - Q_{\psi_1(1)} \tag{40}$$

and

$$M_{(2)}^{22} = Q_{\psi_1(3)} - Q_{\psi_1(2)} \tag{41}$$

Generalized forces $Q_{\psi_1(1)}$, $Q_{\psi_1(2)}$, and $Q_{\psi_1(3)}$ may be obtained using an inverse dynamics approach[83] which involves position, velocity, acceleration, and dynamic equilibrium analyses.

The above scheme provides the values of the parameterized functions at a given $\psi_1 = \psi_{10}$. This procedure must be repeated to obtain the values of the functions for the entire range of ψ_1. A similar approach may be used to obtain other parameterized functions.[83]

Few remarks concerning this formulation are in order: 1) It may appear that the current method requires three kinematic analyses for the necessary computations in the inverse dynamics, but, truly, it only requires one kinematic analysis[83]; 2) the necessary parameterized functions stated above may also be generated using other schemes, however, this method is advantageous especially when a superelement appears as a black box; and 3) the response time of the system may be decreased by implementing this scheme on a multiprocessor machine.

The computational flow chart of a sequential algorithm resulting from these discussions is shown in Figure 7. From this figure, it is clear that the chart may be divided into two major groups (or modules): the superelement function generation group (Group A), and the integration group (Group C). The superelement functions may be generated, 1) in the main program as the integration proceeds, in which case Group A succeeds Group C, or 2) in the main program before the integration begins, in which case Group A precedes Group C, or 3) in a preprocessor, in which case these two groups may exist altogether in two different programs. These programs may be executed on two different processors or on two different machines at different remote sites. The first approach is advantageous when the values of the functions need to be computed only a few times, and the second and third approaches are advantageous when the values of the functions need to be computed a number of times. Note that these functions need to be generated only once, and the same functions may be used in subsequent sets of loadings or various vehicle operating environments.

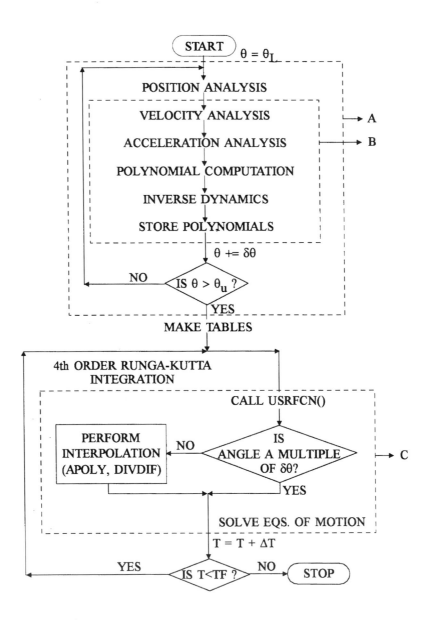

Figure 7. Superelement model based sequential flow graph for vehicle dynamics.

It should also be noted that several tasks in group A follow sequential steps that can be pipelined. For this reason, a portion of group A is represented as group B (Figure 7). The identification of these steps and computational modules play a significant role in the parallel algorithm to be presented next.

4.9. PARALLEL ALGORITHMS

One of the most prominent advantages of using multiprocessor machines and multi-threading operating systems is the achievement of speed up in the execution of computationally intensive programs. However, an effective speedup in such environment usually requires a deep understanding of the structure of the algorithm and the way a proposed parallelized version of the algorithm is affected by the overheads of context switches and shared memory multiprocessor environment. The three paradigms presented below, master-slave, workpile, and pipeline, do not exhaust the ways to design parallel algorithms; one may wish to consider other models as well. The Linda distributed programming system[93] offers a way to synchronize work on multiprocessors or multicomputers. Packages such as PVM[94], MPI[95], and Molecule[96] are tools for writing message-passing applications.

Master-Slave Algorithm

The simplest paradigm is the master-slave. The main master thread launches a set of slave processes on a multiprocessor machine or threads on a multithreading Operating System, and allocates to each slave a portion of the work to be done. The amount of work is known in advance by the master and divided among the slaves. The master starts the slaves and waits for all of them to reach the synchronization point, or barrier. The master may then compute and release the slaves if needed. This pattern is repeated until the work is done.

This is a simple paradigm and is used when the amount of work to be done is known in advance and when it is easy to partition the work into n roughly equal parts that don't depend on each other. The most common application is to a loop, where m iterations of the loop are partitioned among n processes/threads. Usually m is greater than n, so each process/thread invocation executes quite a few iterations of the loop.

The key element of the master-slave paradigm is that every slave process/thread must complete its tasks before the computation can proceed. Rather than requiring each process/thread to exit in order to synchronize one can allow a pool of processes/threads to remain ready to work on subsequent computations as shown in Figure 8. Note that each slave must synchronize to start the computation and to end it.

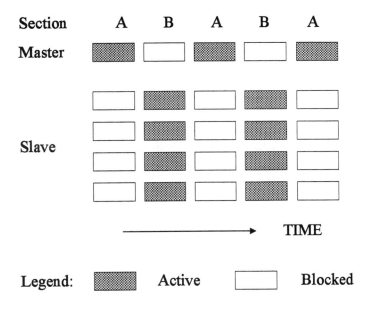

Figure 8. Master-Slave with slave pool.

Workpile Algorithm

Another popular algorithm is the workpile, in which a set of worker processes or threads request chunks of work to do from a "workpile", usually some form of queue. As part of the work, a worker may cause additional entries to be added to the workpile. The pattern terminates when the workpile is finally emptied; alternatively, a worker processor thread may choose to terminate the process.

```
Worker
    while(ptr = work_get(workpile) != NULL) {
        – get work from workpile compute on work assignment indicated
        by "ptr"
        can call work_put(workpile, ptr_new) to add work to the pile
    }
```

In the above template, the worker process/threads loop continuously removing and processing work from the workpile. Work is represented by an application-dependent structure that tells the worker what to do. In the process of performing its assigned task a worker may add new work to the pile.

Pipeline Algorithm

The pipeline is one of the simplest and most compelling paradigms for creating parallel applications. The simple producer-consumer idiom is an example of two-stage pipeline: the producer creates or obtains a block of data and puts a pointer to the data on a queue; the consumer obtains entries from the queue in the same order created by the producer and processes the data. The queue handles all of the synchronization needs of these two processes. When the producer tries to add an entry to a full queue, the producer will block until space in the queue is available. On the other hand, when the consumer tries to obtain an entry from the empty queue, the consumer will block until the producer puts at least one entry in the queue. In this paradigm, a task is passed to a succession of processes/threads, each of which performs part of the overall work required; the processes/threads are much like workers on an assembly line. In most cases, the processing in each pipeline stage is different, but there are applications for homogeneous pipelines in which the code executed in each is the same.

The present research makes use of Master-Slave and Pipeline paradigms to achieve an speedup.

The goal of writing a parallel algorithm is to obtain a good performance on a multiprocessor. Lock contention and Synchronization overheads are some of the performance problem areas. In most cases the overhead should remain fixed or increase more slowly than the computation required to solve the problem. Hence, big problems show better speedups than small problems. Also, a high degree of independent modules result in lesser lock contention and hence a better speedup. Also, a higher grain size (amount of work done by a process) can reduce the synchronization, process initialization, and context switch overheads. Computational load associated with a module must be large enough to overcome this overhead. Depending on the computer architecture, this overhead can significantly differ, and an algorithm designed without considering this overhead may perform poorly. An approach to this problem is to identify subprograms (subroutines, functions, or procedures) which account for most of the program's CPU time. This can be accomplished by two means: First, by carefully counting the number of computations associated with each module, or second, by using the utilities available in the computers. In this study, the DYNIX gprof utility was used to create a profile of the sequential program and to find the above information. The kinematic configuration of the system was also considered to further reduce the computational time.

Following the above study, a parallel flow graph proposed here for the planar vehicle dynamics is shown in Figure 9. The entire computation consists of two major parallel branches of which each represents the computation flow sequence of a superelement (or a suspension systems). Each branch is divided into three major stages, namely, superelement function generation (boxes A and B), polynomial interpolation (boxes C and D), and generalized force computations (boxes E and F). There also exists some parallelism within the computation steps of box A and box B. This parallelism is exploited in the discussion to follow.

Implementation of Master-Slave Algorithm

The discussion above and the flow diagram in Figure 9 suggest the use of a master-slave scheme for the vehicle model. The master/parent process creates a pool of slave/child processes at the beginning of the program. The slaves stay in the pool untill the master process detects work. A slave is then released to perform that work. The slave process, upon completion of its job, goes back to the pool.

Two models, namely, QUEUE and LOCK are available for creating and releasing the slaves. In this study, the LOCK model was implemented because it is simpler and more efficient than the QUEUE model. Figures 10 and 11 respectively, show the LOCK model algorithms for creating and releasing a slave processor. The underlying idea here is to use the UNIX fork() system call to create slave processors. S_lock(&spin_lock) execution by the created slaves results in the generation of the pool of processors vying for the lock (Figure 10). The child tries to obtain the spin_lock but it fails if the master is holding the spin_lock. Hence, the child keeps spinning for the lock.

When additional processors are needed for parallel computations, the master executes a s_unlock call. A s_unlock(&spin_lock) operation by the parent releases the spin_lock which may then be accessed by one of the child processors (Figure 11). The slave that has obtained the spin_lock can release another slave from the pool if desired. However, it can do so only after obtaining an access to a count_lock In this case synchronization among processors is necessary. For the detailed discussion on the above implementation, readers are referred to[97].

4.9.1. Polynomial Computation in Parallel

The following discussion concentrates on the nature of the parallelism within box A of Figure 9. Box B in the same flow diagram has the similar parallelism. Figure 12 shows the nature of dependency of the program shown in box A. SEQ CODE box in this figure consists of velocity and acceleration analyses, polynomial computation, inverse dynamics, and storage of polynomials as shown in box B of Figure 7.

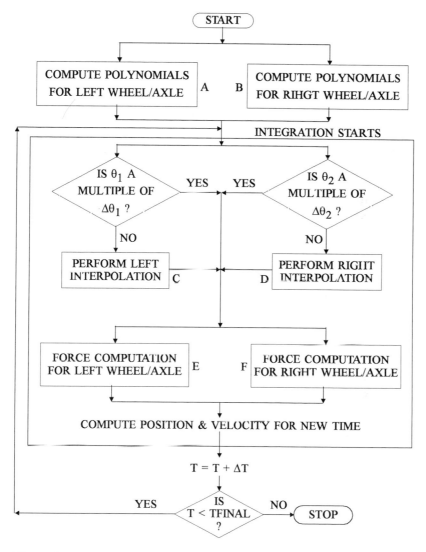

Figure 9. Superelement model based parallel flow graph for vehicle dynamics.

The above scheme leads to a two stage pipeline model. A multistage pipeline model may be obtained first by dividing box A of Figure 7 into several modules such as position, velocity, acceleration, etc., and then putting them in a sequence. Since position analysis takes few iterations, and velocity, accelerations, etc., only take one iteration, a two stage pipeline model is considered appropriate for this study.

```
s_lock(&spin_lock);
   for (slave=0; slave < SLAVE ; ++ slave)
   {        k = slave;
            pid = fork();
                  if (pid !=0) continue;
                  else {        child_id[k] = getpid();
                                s_lock(&spin_lock);
                        }
   }
```

Figure 10. Creation of a pool of slaves - LOCK model

```
      if(pid != 0)                    /* parent or master */
            s_unlock(&spin_lock);
      if(child_id[1] == getpid())     /* child or slave */
      {    s_lock(&count_lock);
           ++count;
           if(count < SLAVE)s_unlock(&spin_lock);          release of
slaves */
           s_unlock(&count_lock);
           ......
           ......
           WORK
           ......
           done_1 = 1;        }

      if(pid !=0)          /* parent waits for children */
            while(done_1 == 0 || ....... )
               ;
```

Figure 11. Release of slaves - LOCK model.

The sequence of the computation may be described as follows: Array 'ay' stores the initial coordinates of the mechanism. At the beginning of the program the position analysis starts with the initial angle $\theta = \theta_L$, where θ_L is the lower limit of θ. The output of this function call (vector 'ay') can allow two concurrent steps to occur and accordingly two processors may be used. Processor one can execute the SEQ CODE for the old angle θ, whereas processor two can call the function position(ay) to compute the vector 'ay' for the new angle $\theta = \theta + \delta\theta$. The above steps are repeated until θ reaches

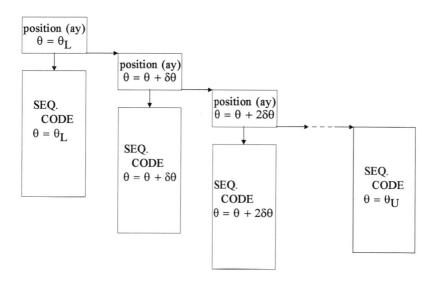

Figure 12. A special pipeline model.

the upper limit $(= \theta_U)$. The execution of the SEQ CODE lags behind the position analysis by at least a step. This is because the SEQ CODE for the angle(θ) can be executed only if the position analysis for that angle has been completed. Hence, for computational efficiency, the two modules in the code must be synchronized.

(a) A Simple Master-Slave Pipeline Model: Figure 13 shows the implementation algorithm for the simple master-slave pipeline model. Since ay[ ] is a shared array, each fork() call results in allowing the new child access the array ay[ ] data. The child proceeds to execute the SEQ CODE which requires the existing values of the ay[ ] vector, whereas the parent process concurrently calls function position(ay) to compute the new values of ay[ ] for the new angle $\theta = \theta + \delta\theta$. The child dies upon the completion of its work (exit(0) call), whereas the parent goes ahead and creates a new child. This results in system overhead. If some parallel work appears later in the program, then new children are created which results in an additional overhead. Table 3 gives an indication of the system overhead due to the varying number of UNIX fork()s and exit()s calls.

Numerical experiments exhibit that the granularity of the position analysis and the SEQ CODE are almost the same, and hence the processors are evenly loaded. The very nature of the problem does not

```
position(ay);
  for (k=1; k<=73; ++k)
    {
      id=fork();
      if((id !=0) && (θ < θ_U))
        {
          θ = θ + δθ ;
          position(ay);
        }
      else
        {
          SEQENTIAL CODE
          . . . . . .
          exit(0);
        }
    }
  wait(0);
  . . . . . .
  . . . . . .
```

Figure 13. An algoritms to implement a pipeline model.

allow execution of more than 2 processors inside the code at any instant, and hence a linear speedup of 2.0 is an ideal expectation. As observed earlier, the extra overhead makes this technique inefficient. The number of fork()s and exit()s can be reduced if a pool of processors are formed initially, and once a child completes its work, it goes back to the pool, thereby being available for the next work.

Table 3. Overhead due to fork()s and exit()s at system load of 0.06

number	CPU time (sec.)
1	0.05
5	0.13
10	0.25
20	0.50
25	0.70
50	1.45
75	2.20

```
if (child_id[0] == getpid()) ay_consumed = 1;
   if (child_id[1] == getpid()) poss_done = 0;
   while (( θ₁ <= θ_U ) && (θ₂ <= θ_U ))
   {          if (child_id[0] == getpid())
              {          while (ay_consumed == 0);
                         ay_consumed = 0;
                         θ₁ = θ₁ + Δθ₁
                         position(ay);
                         poss_done = 1;
              }
              if (child_id[1] == getpid())
              {          while (poss_done == 0);
                         possdone = 0;
                         θ₂ = θ₁
                         ay_consumed = 1;
                         SEQ CODE
                                   }
   }
```

Figure 14. A modified pipeline algorithm.

(b) A Modified Pipeline Model: Figure 14 gives the C code used to simulate a special pipeline model. In this model, the two children are allowed to work on the above pipeline model. A child, depending upon its ID number, initializes the shared variable 'ay_consumed' and 'poss done' to zero(0) or one(1). The variable 'ay_consumed' equal to one causes the first child to go in an infinite loop, and the variable 'poss_done' equal to zero does the same to the second child. Initially the variable ay_consumed is set to one, and the variable poss_done is set to zero. This causes the first (child_id[0]) to proceed and make the position(ay) function call and the second child (child_id[1]) to trap in an infinite while loop. Next, child_id[0] changes the 'poss_done' to one(1), thereby, releasing the child_id[1] from the infinite while loop. The first child is now trapped in the infinite while loop since 'ay_consumed' is zero. Only after the second child has made 'ay_consumed' to one(1), the first child is able to perform the position(ay) call for the new configuration. The above steps are repeated until the limiting case is reached. Thus, a careful usage of 'ay_consumed' and 'poss_done' simulates the special pipeline model.

4.9.2. Integration and Interpolation in parallel

The DYNIX gprof utility indicates that the numerical integration and interpolation consume almost 48% of the total CPU time. Therefore, computer responses can be reduced by exploiting parallelism at this stage. In this study, a fourth order Runge-Kutta method is used to integrate the differential equations. This method calls a function usrfcn() four times per cycle. These are highly sequential steps and may not be performed in parallel. However, parallelism exists within the usrfcn() function calls.

Several functions such as m_fork(), m_single(), m_multi(), m_next(), etc. of the Sequent multiprocessing library are used to simplify the algorithm and achieve computational efficiency. The DYNIX multitasking m_fork() is used to create child processes. The function usrfcn() has 15% of sequential code and hence m_single() is used to execute this portion of the code. m_single() allows only one processor to execute till m_multi() is called. m_next() gives a new value of count to each child and a child depending upon its id (= m_get_myid()) executes a portion of the code. This scheme allows the children to be alive and available for parallel work after the SEQ CODE has been executed by one of the children. After the work has been completed, m_kill_procs() may be used to kill all the children. Although this improves the response time, the presence of 15 % of sequential code prevents the parallel implementation from achieving a linear speedup. Figure 15 gives three algorithms to determine the overhead due to DYNIX multiprocessor utilities. The overhead associated with the three algorithms for varying circumstances is given in Table 4.

Figure 16 shows the code used to implement the integration and interpolation technique. In this study, four slaves were released from the pool of slaves by the parent to perform the above task. Since each of these interpolation stages is independent and the granularity of the work is the same, the static

Table 4. Overhead due to m_fork(), m_single() and m_kill_procs()

number	Case 1	Case 2	Case3
1	0.24	0.31	0.24
5	0.24	0.31	1.61
10	0.24	0.31	3.11
20	0.25	0.32	6.21
30	0.25	0.33	9.31
50	0.26	0.34	15.53
70	0.26	0.34	—
90	0.27	0.35	—
100	0.27	0.35	—
500	0.39	0.53	156.45

```
Case 1:    for (k=0; k<number; ++k)
               m_fork(iter);
           m_kill_procs();

           void iter()
           { · · · · · ·
             · · · · · ·
             · · · · · ·
           }

Case 2:    for (k=0; k<number; ++k
               m_fork(iter);
           m_kill_procs();

           void iter()
           { · · · · · ·
             · · · · · ·
               m_single();
             · · · · · ·
           }

Case 3:    for (k=0; k<number; ++k)
           {
               m_fork(iter);
               m_kill_procs();
           }
```

Figure 15. Algorithms to determine overheads due to DYNIX multiprocessor utilities.

data partitioning scheme was implemented. The polynomial interpolation work was statically divided equally among the children. Computation of some of the terms of the 2-D matrix, which is later solved for acceleration coefficients, is dependent. Therefore, one must take certain precautions while allotting the work to the children so that execution of all the children processes are done independently. The parent performs the equation solution stage at the end of each usrfcn() call and advances the response to the next time step.

```
rkmethod()
  {
      usrfcn();
          if(pid != 0)          /* parent */
             SEQ CODE
      usrfcn();
          if(pid != 0)
             SEQ CODE
      usrfcn();
          if(pid != 0)
             SEQ CODE
      usrfcn();
          if(pid != 0)
             SEQ CODE
  }
  void usrfcn()
  {


      if(pid != 0)          /* parent */
          SEQ WORK

    /* assign work to different children */

      if(n_child_id[0] == getpid())
      {     s_lock(&n_count_lock);
            ++n_count;
            if(n_count < SLAVE)s_unlock(&next_spin_lock);
            s_unlock(&n_count_lock);

            ...........
            ...........
            WORK
            ...........
            done_1 = 1;
      }


   if(pid !=0)      /* parent */
      {     while(done_1 == 0 || .... )      /* wait for children */
            ;
         /* call equation solver */
      }
```

Figure 16. A parallel code for integration and polynomial interpolation.

Table 5. Initial configuration of planar vehicle model

Body no.	Initial body coordinates		
no.	X(m)	Y(m)	(rad)
1	1.5	0.575	0.0
2	5.85	0.575	0.0
3	3.675	0.975	0.0
4	2.116	0.725	0.464
5	5.234	0.725	-0.464
6	1.066	0.909	-0.154
7	6.284	0.909	0.154

4.10. NUMERICAL RESULTS

In order to demonstrate the feasibility and efficiency of the present scheme, dynamics of the planar vehicle system was simulated on a Sequent Balance 8000 multiprocessing computer using three schemes, namely, the Cartesian[36], the sequential superelement[83], and the parallel superelement presented here. The kinematic configuration and the properties of the vehicle are given in a previous section. The results of the Cartesian model is included in this paper in order to compare the results of this method with a well established classical method.

In the Cartesian model, each link of the vehicle is treated as an element. However, in the superelement models, the Chassis is treated as a Cartesian element and the suspension systems are treated as superelements. The equilibrium configuration of the vehicle is obtained by allowing it to move freely on a flat terrain. The initial configuration of the vehicle for this simulation is given in Table 5. The damping response of the vehicle due to nonequilibrium condition is shown in Figure 17.[97]

In a subsequent study, the dynamics of the vehicle moving over a bump (Figure 18) at a speed of 2.75 m/sec is considered. In order to simulate this condition, the x coordinate of the chassis is kept fixed and the ground is given a negative velocity of 2.75 m/sec. Equilibrium configuration obtained in the previous simulation is taken as the initial condition for this simulation. The vertical response of the chassis for this case is shown in Figure 19.[83]

For the second simulation, the sequential Cartesian model took 2937 seconds of CPU time. Tables 6 and 7 show the CPU time taken at various stages by sequential and parallel superelement models. Table 8 shows the speed-up achieved by the parallel over the sequential superelement model. These results show that for the planar vehicle model considered here the new parallel algorithm is approximately 3.65 and 18.91 times faster

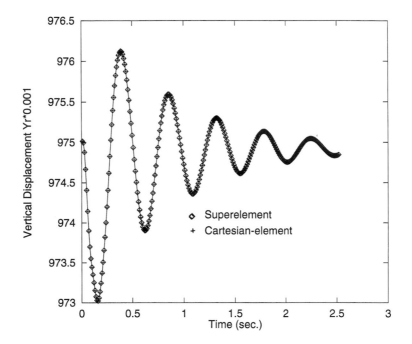

Figure 17. Vertical response of the vehicle on a flat terrain.

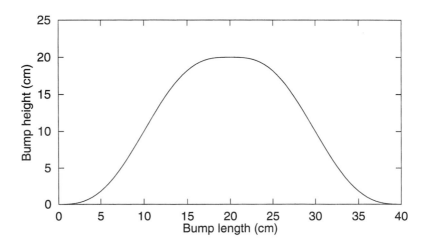

Figure 18. Bump profile.

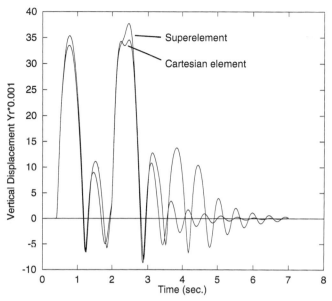

Figure 19. Vertical response of the vehicle over a bump.

than the sequential superelement algorithm[83] and the sequential Cartesian algorithm[36], respectively.

4.11. CONCLUSIONS

A brief review of schemes for vehicle dynamics and multibody dynamics has been presented. Three models, a Cartesian element, a Lagrangian element, and a superelement have been discussed for a planar vehicle dynamics. The Cartesian element and the superelement models have been extended for multiprocessor computers. The algorithm uses an inverse dynamics based approach for generating superelement parameters. Techniques for determining the segments of the algorithm that can be executed in parallel have been discussed. Parallel algorithms to simulate the pipeline model were also designed and implemented on a Sequent Balance 8000 multiprocessing computer. The results show that the superelement model based parallel algorithms can significantly reduce the multi-body dynamics response time. The algorithm presented is general and may be used in various other applications. Future reasearch could venture into determining the amount of speedup that can be achieved if the superelement modeling is performed on a multi-threaded operating system running on a multi-processor hardware.

Table 6. CPU time for dynamic simulation of planar vehicle (sequential superelement algorithm)

Stage description	CPU secs	Cumulative secs
Seq. execution of polynomials for left wheel	16.15	16.15
Seq. execution of polynomials for right wheel	17.04	33.19
Seq. execution of poly interpolation for left and right wheel-axle assembly and integration of the diff. equations	535	568.19

Table 7. CPU time for dynamic simulation of planar vehicle (parallel superelement algorithm)

Stage description	CPU secs	# of procs.	Cumulative secs
Execution of polynomials for left wheel (2 processors) is 9.13 Execution of polynomials for right wheel(2 processors) is 9.31	9.31	4	9.31
Execution of poly interpolation for left and right wheel-axle assembly and integration of the diff. equations	146	4	155.31

Table 8. Speed-ups achieved by parallel superelement model

Stage description	# of procs	Speed-up achieved
Computation of polynomials for left wheel-axle assy.	2	1.77
Computation of polynomials for right wheel-axle assy.	2	1.83
Polynomial interpolation and for left and right wheel-axle assembly and integration of the diff. equations	4	3.66

References

1. Zeid, A. and D. Chang, 1991, "Simulation of multibody systems for the computer aided design of vehicle dynamic controls," *Vehicle Dynamics and Electronic Controlled Suspensions SAE Special Publications no. 861*, Publ. by SAE, Warrendale, PA, USA. pp. 63–70.
2. Sohoni, V.N., L.J. Duchnowski and J.R. Winkelman, 1990, "Multi-body modeling of suspension kinematics for control design." *Proceedings of the American Control Conference.* Publ. by American Automatic Control Council, Green Valley, AZ, USA (IEEE cat. no. 90CH2896-9). pp. 1375–1380.

3. Gim, G. and P.E. Nikravesh, 1991, "Comprehensive three dimensional models for vehicle dynamic simulations." *Proceedings of the 6th International Pacific Conference on Automotive Engineering Proc 6 Int Pac Conf Autom Eng.* Publ by Korea Soc of Automotive Engineers, Inc, 1638–3, Socho-dong, Seoul, South Korea. pp. 1089–1097.

4. Crolla, D.A. and H. Schwanghart, 1992, "Vehicle dynamics. Steering I." *Journal of Terramechanics*, **29**(1), 7–17.

5. Zeid, A. and D. Chang, 1991, "Simulation of multibody systems for the computer -aided design of vehicle dynamic controls." SAE Technical Paper Series. Publ by SAE, Warrendale, PA, USA, 910236. 8p.

6. Pereira, M.S. and P.I. Proenca, 1991, "Dynamic analysis of spatial flexible multibody systems using joint co-ordinates." *International Journal for Numerical Methods in Engineering*, **32**(8), 1799–1812.

7. Lee, J.N. and P.E. Nikravesh, 1994, "Steady-state analysis of multibody systems with reference to vehicle dynamics." *Nonlinear Dynamics*, **5**(2), 181–192.

8. Freeman, J.S. and S.A. Velinsky, 1993, "Mixed planar and spatial modeling for multibody dynamics." *Advances in Design Automation American Society of Mechanical Engineers, Design Engineering Division (Publication) DE*, **65**(1). Publ. by ASME, New York, NY, USA. pp. 37–42.

9. Kortuem, W. and R.S. Sharp, 1991, "Report on the state-of-affairs on application of multibody computer codes to vehicle system dynamics," *Vehicle System Dynamics*, **20**(3-4), 177–184.

10. Kortuem, W. and R.S. Sharp, 1992, "IAVSD review of multibody computer codes for vehicle system dynamics." *Transportation Systems – 1992 American Society of Mechanical Engineers, Dynamic Systems and Control Division (Publication) DSC*, **44**. Publ. by ASME, New York, USA. pp. 1–14.

11. Lieh, J. and I. Haque, 1990, "Symbolic dynamics and control analysis of elastic vehicles with active suspensions." *Transportation Systems 1990 American Society of Mechanical Engineers, Applied Mechanics Division, AMD*, **108**. Publ. by ASME, New York, NY, USA. pp. 237–245.

12. Friberg, O., 1991, "Method for selecting deformation modes in flexible multibody dynamics." *International Journal for Numerical Methods in Engineering*, **32**(8), 1637–1655.

13. Lieh, J. and I. Haque, 1991, "Closed-form formalism for controlled and hybrid rigid/ elastic multibody systems. Part II. Symbolic implementation and applications." *Advances in Design Automation American Society of Mechanical Engineers, Design Engineering Division (Publication) DE*, **32**(2). Publ. by ASME, New York, NY, USA. pp. 411–416.

14. Shabana, A.A., 1990, "Computer aided analysis of flexible multibody vehicle systems." SAE (Society of Automotive Engineers) *Transactions*, **99**(Sect. 5), 969–980.

15. Lieh, J., 1994, "Separated-form equations of motion of controlled flexible multibody systems." *Journal of Dynamic Systems, Measurement and Control, Transactions of the ASME*, **116**(4), 702–712.

16. Nikravesh, P.E., J.A.C. Ambrosio, and M.S. Pereira, 1990, "Rollover simulation and crashworthiness analysis of trucks," *Forensic Engineering*, **2**(3), 387–401.

17. Lupker, H.A., P.J.A. de Coo, J.J. Nieboer and J. Wismans, 1991, "Advance in MADYMO crash simulations." *Side Impact Occupant Protection Technologies SAE Special Publications 851*. Publ. by SAEM Warrendale, PA, USA. pp. 135– 146.

18. Gupta, A.D., 1990, "Evaluation of drag loading models in overturning response codes." *Proc .1990 ASME Int. Comput. Eng. Conf. Expo.* Publ. by ASCE, Boston Society of Civil Engineers Sect., Boston, MA, USA. pp. 57–65.

19. Boneill, C. and J.P. Mizzi, 1993, "Identification of parameters for a vehicle submitted to a crash." *Dynamics and Vibration of Time-Varying Systems and Structures American Society of Mechanical Engineers, Design Engineering Division (Publication) DE*, **56**. Publ. by ASME, New York, NY, USA. pp. 245–257.

20. Dias, J.P. and M.S. Pereira, 1994, "Design for vehicle crashworthiness using multibody dynamics." *International Journal of Vehicle Design*, **15**(6), 563–577.

21. Ma, D. and H. Lankarani, 1994, "Multibody/finite element analysis approach for modeling of crash dynamic responses." *20th Design Automation Conference American Society of Mechanical Engineers, Design Engineering Division (Publication) DE*, **69**(1). ASME, New York, NY, USA. pp. 55–64.

22. Sayers, M.W. and C.W. Mousseau, 1990, "Real-time vehicle dynamic simulation obtained with a symbolic multibody program." *Transportation Systems 1990 American Society of Mechanical Engineers, Applied Mechanics Division, AMD*, **108**. Publ. by ASME, New York, NY, USA. pp. 51–58.

23. Mousseau, C.W., M.W. Sayers and D.J. Fagan, 1991, "Symbolic quasi-static and dynamic analyses of complex automobile models." *Vehicle System Dynamics*, **20**(Suppl.), 446–459.

24. Sayers, M.W. and P.S. Fancher, 1993, "Hierarchy of symbolic computer-generated real-time vehicle dynamics models." *Transportation Research Record*, **1403**, 88–97.

25. Kortum, W. 1990, "Analysis of railway vehicle system dynamics with the multibody program MEDYNA." *IEEE Technical Papers Presented at the Joint ASME/IEEE/AAR Railroad Conference* (Association of American Railroads). Publ. by IEEE, IEEE Service Center, Piscataway, NJ, USA (IEEE cat. no. 90CH2947-0). pp. 57–63.

26. Pascal, J.P., G. Sauvage and P. Delfosse, 1993, "Use of multibody software in the wheel/rail contact as applied to very high speed tests". *High Speed Ground Transportation Systems I.* Proc. First Int. Conf. High Speed Ground Transp. Syst. 1993. Publ. by ASCE, New York, NY, USA. pp. 497–510.

27. Yang, G. and A.D. de pater, 1991, "Determination of the nonlinear motion of a railway vehicle." *Vehicle System Dynamics*, **20**(Suppl.), 225–239.

28. Fisette, P. and J.C. Samin, 1991, "Lateral dynamics of a light railway vehicle with independent wheels." *Vehicle System Dynamics*, **20**(Suppl.), 157–171.

29. Diomin, Y.V., 1994, "Stabilization of high-speed railway vehicles." *Vehicle System Dynamics*, **23**(2), 107–114.

30. Diomin, Y.V., E.N. Kovtun and O.M. Markova, 1994, "Self-excited vibrations of railway vehicle with dry friction units." *Vehicle System Dynamics*, **23**(1), 71–83.

31. Yang, G., 1994, "Aspects in modelling a railway vehicle on an arbitrary track." *Rail Transportation American Society of Mechanical Engineers, Rail Transportation Division (Publication) RTD 8 1994*. ASME, New York, NY, USA. pp. 31–36.

32. Watanabe, K., H. Murakami, M. Kitano and T. Katahira, 1993, "Experimental characterization of dynamic soil-track interaction on dry sand" *Journal of Terramechanics*, **30**(2), 111–131.

33. Sarwar, M.K., A.A. Shabana and T. Nakanishi, 1994, "Design methodology for tracked vehicles using experimentally identified modal parameters and the finite element method." *20th Design Automation Conference American Society of Mechanical Engineers, Design Engineering Division (Publication) DE*, **69**(2). ASME, New York, NY, USA. pp. 491–502.

34. Daberkow, A. and W. Schirhlen, 1994, "Concept, development and implementation of DAMOS-C: The object oriented approach," *Computers in Engineering, Proceedings of the International Conference and Exhibit*, **2**, 1994, ASME, New York, NY. pp. 937–951.

35. Orlandea, N., M.A. Chace and D.A. Calahan, 1977, "A Sparsity Oriented Approach to the Dynamic Analysis and Design of Mechanical Systems, Part I and II, *Engng for Ind.*, **99**, 773–784.

36. Wehage, R.A. and E.J. Haug, 1981, "Generalized Coordinate Partitioning for Dimension Reduction in Analysis of Constrained Dynamic Systems, *ASME J. Mech. Design*, **104**, 247–255.

37. Kane, T.R. and D.A. Levinson, 1983, "Multi-body Dynamics," *ASME J. Applied Mechanics*, **50**, 1071–1078.

38. Shabana, A.A. and R.A. Wehage, 1984, "Spatial Transient Analysis of Inertia-variant Flexible Mechanical Systems," *ASME J. Mech. Trans. and Auto. in Design*, **106**, 172–178.

39. Mani, N.K., E.J. Haug and K.E. Atkinson, 1985, "Application of Singular Value Decomposition for Analysis of Mechanical System Dynamics," *ASME J. Mech. Trans. and Auto. in Design*, **107**, 82–87.

40. Chang, C.O. and P.E. Nikravesh,, 1985, "An Adaptive Constraint Violation Stabilization Method for Dynamic Analysis of Mechanical Systems," *ASME J. Mech. Trans. and Auto. in Design*, **107**, 488–492.

41. Wampler, C., K. Buffington and J. Shu-hui,, 1985, "Formulation of Equations of Motion for Systems Subject to Constraints," *ASME J. of Appl. Mech.*, **52**, 465–470.

42. Singh, R.P. and P.W. Likins,, 1985, "Singular Value Decomposition of Constrained Dynamical Systems," *ASME J. of Appl. Mech.*, **52**, 943–948.

43. Haug, E.J. and M. McCullough,, 1985, "A Variational-Vector Calculus Approach to Machine Dynamics," *ASME J. Mech. Trans. and Auto. in Design*, **108**, 25–30.

44. Kim, S.S. and M.J. Vanderploeg, 1986, "QR Decomposition for State Space Representation of Constrained Mechanical Dynamic Systems, *ASME J. Mech. Trans. and Auto. in Design*, **108**, 183–188.

45. Liang, C.G. and G.M. Lance, "A Differentiable Null Space Method for Constrained Dynamic Analysis", ASME paper no. 85–DET-86.

46. Jalon, J.G.D., J. Unda, A. Avello and J.M. Jimenez, "Dynamic Analysis of Three-Dimensional Mechanisms in Natural Coordinates," ASME paper no. 86–DET 137.

47. Agrawal, O.P. and A.A. Shabana, 1986, "Application of Deformable body Mean-axis to Flexible Multi-body System Dynamics," *Comp. Meth. in Appl. Mech. and Engng.* **56**, 217–245.

48. Li, T.W. and G.C. Andrews, 1987, "Application of the Vector Network Method to Constrained Mechanical Systems," *ASME J. Mech. Trans. and Auto. in Design*, **108**, 471–480.

49. Brandl, H., R. Johanni and M. Otter, 1987, "An Algorithm for the Simulation of Multibody Systems with Kinematic Loops," *Proceedings of the Seventh World Congress on the Theory of Machines and Mechanisms*, **1**, 407–411.

50. Hiller, M. and A. Kecskemethy, 1987, "A Computer-Oriented Approach for the Automatic Generation and Solution of the Equations of Motion for Complex Mechanisms," *Proceedings of the Seventh World Congress on the Theory of Machines and Mechanisms*, **1**, 425–430.

51. Ider, S.K. and F.M.L. Amirouche, 1988, "Coordinate Reduction in Constrained Spatial Dynamic Systems - A New Approach," *J. Appl. Mech.* **52**, 899–905.

52. Agrawal, O.P. and S. Saigal, 1989, "Dynamic Analysis of Multi-body Systems Using Tangent Coordinates," *Computers and Structures*, **31**(3), 349–355.

53. Roberson, R.E. 1977, "Computer Oriented Dynamic Modeling of Spacecraft: Historical Evolution of Eulerian Multi-body Formalisms Since 1750," *Proc. 28th Int. Astronaut. Congr. Prag*, IAF Paper 77–A11.

54. Chace, M.A. and P.N. Sheth, 1973, "Adaptation of Computer Techniques to the Design of Mechanical Dynamic Machinery," ASME Paper 73–DET-58, ASME, NY.

55. Paul, B.,, 1975, "Analytical Dynamics of Mechanisms – A Computer Oriented Overview," *Mechanism and Machine Theory*, **10**, 481–507.

56. Jerkovsky, W., 1978, "The Structure of Multi-body Dynamics Equations," *J. Guidance and Control*, **1**, 173–182.

57. Likins, P.W., 1975, "Point-connected Rigid Bodies in a Topological Tree," *Celestial Mechanics*, **11**, 301–317.

58. Kane, T.R. and A.D. Levinson, 1980, "Formulation of Equations of Motion for Complex Spacecraft," *J. Guidance Control*, **3**, 99–112.

59. Schwertassek, R. and R.E. Roberson,, 1985, "A Perspective on Computer- Oriented Multibody Dynamical Formalisms and their Implementations," *Dynamics of Multibody Systems*, ed. G. Bianchi and W. Schiehlen, Springer-Verlag, New York, pp. 261–274.

60. Huston, R.L., 1981, "Multi-body Dynamics Including the Effects of Flexibility and Compliance,"*Computers and Structures* **14**, 443–451.

61. Schiehlen, W., 1990, *Multibody Systems Handbook*, New York, NY.

62. Richard, M.J. and C.M. Gosselin, 1993, "A survey of simulation programs for the analysis of mechanical systems," *Mathematics and Computers in Simulation*, **35**, 103–121.

63. Wittenburg, J., 1977, *Dynamics of Systems of Rigid Bodies*, Teubner, Stuttgart.

64. Nikravesh, P.E., 1988, *Computer Aided Analysis of Mechanical Systems*, Prentice Hall, New Jersey.
65. Haug, E.J., 1989, *Computer Aided Kinematics and Dynamics of Mechanical Systems, Volume I: Basic Methods*, Allyn and Bacon, Massachusetts.
66. Shabana, A.A. 1989, *Dynamics of Multibody Systems*, Prentice Hall, New Jersey.
67. Huston, R.L., 1990, *Multibody Dynamics*, Butterworth-Heinemann, Stoneham.
68. Amirouche, F.M.L., 1992, *Computational Methods in Multibody Dynamics*, Prentice Hall, New Jersey.
69. Li, C., 1989, "A New Lagrangian Formulation of Dynamics for Robot Manipulators," *ASME J. Dynamic Systems, Measurement, and Control*, **111**, 559–567.
70. Featherstone, W.R. 1983, "The Calculation of Robot Dynamics using Articulated-Body Inertias," *Robotics Research*, **2**(1), 13–30.
71. Bae, D. and E.J. Haug, 1987, "A Recursive Formulation For Constrained mechanical systems, Part I. Open loop systems," *Mech. Struct. Mach.*, **15**(3), 359–382.
72. Wehage, R.A. 1989, "Solution of Multibody Dynamics using Natural Factors and Iterative Refinement – Part I: Open Kinematic Loops," *Advances in Design Automation – 1989*, **DE-19–3**, 125–132.
73. Hiller, M. and A. Kecskemethy, 1987, "A Computer-Oriented Approach for the Automatic Generation and Solution of the Equations of Motion for Complex Mechanisms," *The Theory of Machines and Mechanisms*, Proc. of the 7th world congress, **1**, 425–430.
74. Bae, D. and E.J. Haug, 1987, "A recursive formulation for constrained mechanical systems, Part II. Closed loop systems," *Mech. Struct. Mach.*, **15**(4), 481–506.
75. Wehage, R.A., 1989, "Solution of Multibody Dynamics using Natural Factors and Iterative Refinement – Part II: Closed Kinematic Loops," *Advances in Design Automation – 1989*, **DE-19–3**, 133–139.
76. Tsai, F. and E.J. Haug, 1991, "Real-time multibody system dynamic simulation: Part I. A modified recursive formulation and topological analysis," *Mech. Struct. Mach.*, **19**(1), 99–127.
77. Rodriguez, G., A. Jain, and K. Kreutz-Delgado, 1992, "Spatial operator algebra for multibody system dynamics," *J. of the Astronautical Sciences*, **40**(1), 27–50.
78. Wehage, R.A. and A.A. Shabana, 1989, " Application of Generalized Newton-Euler Equations and Recursive Projection Methods to Dynamics of Deformable Multibody Systems," *Advances in Design Automation – 1989*, **DE-19–3**, 17–25.
79. Hiller, M., 1995, "Dynamics of Multiloop Systems," *Kinematics and Dynamics of Multi-Body Systems*, ed. J. Angeles and A. Kecskemethy, Springer Verlag, Wein, NY, pp. 75–165.
80. Hiller, M., 1995, "Multiloop kinematic chains," *Kinematics and Dynamics of Multi-Body Systems*, ed. J. Angeles and A. Kecskemethy, Springer Verlag, Wein, NY, pp. 167–215.
81. Kecskemethy, A., 1995, "Object oriented modelling of mechanical systems," *Kinematics and Dynamics of Multi-Body Systems*, ed. J. Angeles and A. Kecskemethy, Springer Verlag, Wein, NY, pp. 217–276.
82. Agrawal, O.P. and R. Kumar, 1989, "A Superelement Model for Analysis of Multi-body System Dynamics," *Computers and Structures*, **32**, 1081–1091.
83. Agrawal, O.P. and R. Kumar, 1991, "A General Superelement Model on a Moving Reference Frame for Planar Multibody System Dynamics," *J. Vibration and Acoustics*, **113**, 43–49.
84. Agrawal, O.P. and S.L. Chung, 1991, "A Superelement Model Based on Lagrangian Coordinates for Multibody System Dynamics," *Computers and Structures*, **37**, 957–966.
85. Gluck, R., 1986, "Hope for Simulating Flexible Spacecraft," Aerospace America, **24**(11), 40–44.
86. Dubetz, M.W., J.G. Kuhl and E.J. Haug, 1988, "A network implementation of real-time dynamic simulation with interactive animated graphics," *Advances in Design Automation*. ASME Des. Eng. Div. Publ. **14**, 519–523.
87. Tak, T. and S.S. Kim, 1989, "Design Sensitivity Analysis of Multibody Dynamic Systems for Parallel Processing," *Advances in Design Automation - 1989*, **DE-19–3**, 9–16.
88. Tsai, F. and E.J. Haug, 1991, "Real-time multibody system dynamic simulation: Part II. A parallel algorithm and numerical results," *Mech. Struct. Mach.*, **19**(2), 129–162.

89. Bae, D.S. and E.J. Haug, 1988, "A recursive formulation for constrained mechanical systems, Part III. Parallel Processor Implementation," *Mech. Struct. Mach.*, **16**(2), 249–269.
90. Chang, J.L. and S.S. Kim, 1989, "A low-cost real-time man-in-the-loop simulation for multibody systems," *Advances in Design Automation - 1989*, **DE-19–3**, 95–99.
91. Kurdila, A.J. and M.P. Kamat, 1990, "Concurrent multiprocessing for calculating nullspace and range space bases for multibody simulation," *AIAA Journal*, **28**(7), 1224–1232.
92. Kumar, R., 1988, "Dynamic Analysis of Multi-body Systems Using Superelement Models," M.S. Thesis, Department of Mechanical Engineering, Southern Illinois University, Carbondale, IL 62901.
93. Carriero, N., D. Gelertner, 1989, *Linda in Context, Communications of the ACM*, **32**(4), 444–458.
94. Geist, A., A. Beguelin, J. Dongarra, W. Jiang, R. Manchek, V. Sunderam, 1994, *PVN: Parallel Virtual Machine*, MIT Press, Cambridge, Mass.
95. Gropp, W., E. Lusk, and A. Skjellum, 1994, *Using MPI: Portable Parallel Programming with the Message-Passing Interface*, MIT Press, Cambridge, Mass.
96. Xu, Z. and K. Hwang, 1989, Molecule: A Language Construct for Layered Development of Parallel Programs, *IEEE Transactions on Software Engineering*, **15**(5), 587–599.
97. Kumar, R., 1990, "A New Parallel Algorithm For Analysis of Vehicle Dynamics," M.S. Thesis, Department of Computer Science, Southern Illinois University, Carbondale, IL 62901.

5 ROBUST CONTROL DESIGN OF A HIGH PERFORMANCE FLEXIBLE ELECTRO-MECHANICAL SYSTEM

YOSSI CHAIT[1], MAARTEN STEINBUCH[2] and MYOUNG SOO PARK[3]

[1]*Mechanical Engineering Department, University of Massachusetts, Amherst, MA 01003, USA*
[2]*Philips Research Laboratories, Prof. Holstlaan 4, 5656 AA Eindhoven, The Netherlands*
[3]*Applied Light, Inc., 3640 Main St., Springfield, MA 01107, USA*

This chapter considers the design and implementation of robust multivariable controllers applied to a Compact Disc mechanism. The design objective is to achieve good track-following and focusing performance in the presence of disturbances and structured uncertainty in the plant dynamics. In particular, the plant dynamics include lightly damped mechanical vibration modes. The robust performance problem has been solved in the μ-framework using the DK-iteration scheme and using a new, inversion-free, multivariable Quantitative Feedback Theory (QFT) technique. Implementation results of both designs in a Digital Signal Processor (DSP) environment are reported. The efficacy of the two design techniques as applied to this class of problems is discussed.

5.1. INTRODUCTION

A common problem encountered in control of large flexible space structures is the complexity of the dynamics, namely, a large number of lightly damped flexible modes in the systems. To meet stringent performance specifications, such structures must be actively controlled so to minimize the contribution of the flexible modes in the transient response. Control design requires knowledge of the plant dynamics, and the more accurate the model is, the better the achievable controlled performance is with lower costs. Analytic

211

modeling of such structures is an inexact science, at the least, while obtaining models using extensive experimentation is often not feasible. Therefore, achieving robustness for parasitic dynamics becomes critical for successful design of high performance control systems.

In contrast to such systems, small actuator drives such as compact disc and hard disc drive mechanisms, are easy to access with experiments. In these high-volume electronics applications, the problem is to allow for tolerances during production, so as to decrease production costs. Hence, again we see the motivation to design robust performance controllers. Similarly to large flexible structures the dynamics includes flexible modes. Although model identification is more feasible, uncertainty modeling is still necessary to account for changes from product to product.

The research reported here focuses on all aspects of the robust control design for this class of systems including modeling of the dynamics, uncertainty representation, application of two leading control design techniques (μ-synthesis and QFT) and implementation.

In the next section we explain the basic operation of a CD-ROM mechanism. Section 5.4 discusses modeling aspects relevant for robust control design. The performance specifications are defined in Section 5.5. Details of two robust control designs using μ-synthesis and QFT techniques are presented in Sections 5.6–5.7, respectively. Section 5.8 provides details of the experimental setup and the implementation results of the two designs. Finally, conclusions and references are found in Sections 5.9–5.10, respectively.

5.2. THE COMPACT DISC MECHANISM

A Compact Disc player is an optical decoding device that reproduces high-quality audio from a digitally coded signal recorded as a spiral shaped track on a reflective disc.[1] Recently, apart from the audio application, other optical data systems (CD-ROM, Optical Data Drive) and combined audio/video and multi-media (CD-Interactive, MMCD) applications have emerged. An important research area for these applications is the possibility to increase the rotational frequency of the disc to obtain faster data readout and shorter access time. For higher rotational speeds, however, a higher servo bandwidth is required that approaches the resonance frequencies of bending and torsional modes of the CD-mechanism. Moreover, the system behavior varies from player to player due to the required manufacturing tolerances of CD players in mass-production, which explains the need for robustness of the controller.

The Compact Disc mechanism (Figure 1) investigated in this study is composed of a turn-table DC-motor for the rotation of the disc, and a balanced

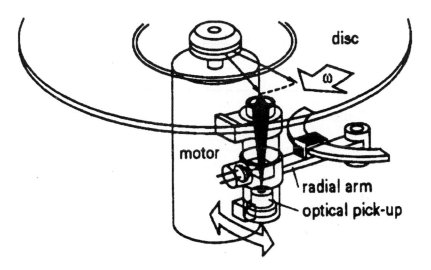

Figure 1. Schematic view of a rotating arm Compact Disc mechanism.

radial arm for track-following. An optical element is mounted at the end of the radial arm. A diode located in this element generates a laser beam that passes through a series of optical lenses to reflect a spot on the information layer of the disc. An objective lens, suspended by two parallel leaf springs, can translate in a vertical direction for focusing action.

Both the radial and the vertical (focus) position of the laser spot, relative to the track of the disc, must be controlled actively. To accomplish this, the controller uses position-error information provided by four photodiodes. As an input to the system, the controller generates the required control currents to the radial and focus actuator (which both are permanent-magnet/coil systems).

In Figure 2, a block-diagram of the control system is shown. The difference between the radial and vertical track physical position and the actual spot position is detected by the optical pick-up; it generates a radial error signal e_{rad} and a focus error signal e_{foc} via the optical gain K_{opt}. The controller $K(s)$ feeds the system with the currents I_{rad} and I_{foc}. The matrix transfer function from control currents to spot position is indicated by $H(s)$. Unique to this class of systems, only the position-error signals (after the optical gain) are available for measurement. Neither the true spot position nor the track position are available as signals.

In current systems the controller $K(s)$ is formed by two separate PID controllers[1,2], thus creating two single-input single-output (SISO) control

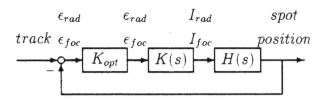

Figure 2. Configuration of the control loop.

loops. This is possible because the dynamic interaction between both loops is relatively low, especially from radial current to focus error. In the CD-ROM mechanism under consideration, the radial loop has a bandwidth of 500 Hz, while the bandwidth for the focus loop is 800 Hz. For more demanding applications, as discussed in Section 5.2, it becomes necessary to consider enhancement of the servo design.

5.4. MODELING

A measured frequency response of the CD-mechanism $G(s) = K_{opt}H(s)$ is given in Figure 3 (magnitude only). At low frequencies, the rigid body mode of the radial arm and the lens-spring system (focus loop) can be easily recognized. It appears as a double integrator in the G_{11} and G_{22} elements of the matrix transfer function (MTF) G. At higher frequencies the measurement shows parasitic dynamics, especially in the radial direction. Experimental modal analyses and finite element calculations have revealed that these phenomena are due to mechanical resonances of the radial arm, mounting plate and disc (flexible bending and torsional modes).

In Figure 4, a typical parasitic mode is shown, calculated with a finite element program.

Using frequency-domain-based system identification, each element of the frequency response has been fitted separately using an output error model structure with a least-square criterion.[3] Frequency dependent weighting functions have been used to improve the accuracy of the fit. The G_{21} element was especially difficult to fit because of the non-proper behavior in the frequency range of interest. In general, the high-frequency response of the off-diagonal elements is not very reliable since the coherence of the measurements was very low.

Combining the separate fits resulted in a 37th order multivariable state-space model. Using frequency weighted balanced reduction[4,5], this model was reduced to a 21st order model without significant loss in accuracy in the servo bandwidth. The frequency response of the model is shown in Figure 3.

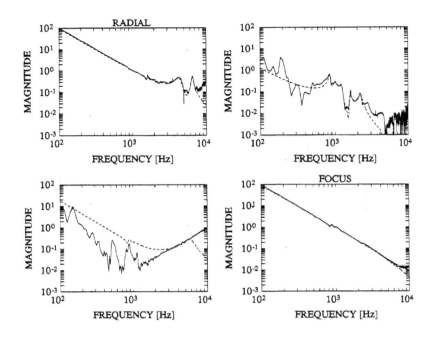

Figure 3. Measured frequency response of the CD-mechanism (- - -) and of the identified 21st order model (- -).

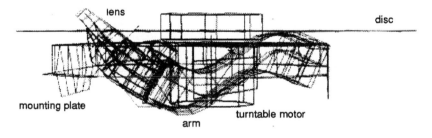

Figure 4. Side view of the 4 kHz bending mode of the radial arm.

5.3.1. Uncertainty Modeling

The most important variations in the CD dynamics from performance view point are:

(i) unstructured difference between model and measurement,

(ii) uncertain interaction,

(iii) uncertain actuator gain, and

(iv) uncertainty in the frequencies of the parasitic resonances.

The first uncertainty stems from the fact that our nominal model only is an approximation of the measured frequency response due to imperfect modeling. Further, a very high-order nominal model is undesirable in some robust control design techniques (e.g., μ-synthesis), since they yield controllers with the state dimension of the nominal model plus weighting functions. For that reason, our nominal model only describes the rigid body dynamics and the resonance modes that are most relevant for control design. Unmodeled high-frequency dynamics and the unstructured difference between model and measurement will be modeled as a complex valued additive perturbation Δa, bounded by a high-pass weighting function.

The remaining uncertainty sources (ii), (iii) and (iv), are all intended to reflect that manufacturing tolerances manifest themselves as variations in the frequency response from one CD mechanism to another. By so doing, we are able to appreciate the consequences of manufacturing tolerances on control design.

The uncertain interaction (ii) is modeled using two scalar output multiplicative parametric perturbations:

$$y = (I + \Delta_o)Cx = Cx + w_o$$
$$W_o = \Delta_o z_o$$
$$z_o = Cx$$

where

$$\Delta_o = \begin{bmatrix} 0 & w_{o1}\delta_{o1} \\ w_{o2}\delta_{o2} & 0 \end{bmatrix}$$

The scalar weights w_{o1} and w_{o2} are chosen equal to 0.1, implying 10% uncertainty. Note that these perturbations also perturb G_{11} and G_{22}. However, this effect is very small in comparison to the perturbation of the off-diagonal elements.

The interaction is assumed to be uncertain since our model describes $K_{\mathrm{opt}}H(s)$. Hence, in control design we can only specify the required disturbance attenuation after K_{opt} (position-error e) and not at the disc itself (position-error ε). When K_{opt} is non-diagonal good disturbance attenuation at e does not imply good disturbance attenuation at the disc. We compensate for this by assuming uncertainty in the amount of interaction in the MTF $G(s)$.

Dual to the uncertain interaction, the uncertain actuator gains (iii) are modeled as two scalar input multiplicative parametric perturbations:

$$\dot{x} = Ax + B(I + \Delta_i)u = Ax + Bw_i$$
$$w_i = \Delta_i z_i$$
$$z_i = u$$

where

$$\Delta_i = \begin{bmatrix} 0 & w_{i1}\delta_{i1} \\ w_{i2}\delta_{i2} & 0 \end{bmatrix}$$

The gain of each actuator is perturbed by 5%. With this value, also non-linear gain variations in the radial loop due to the rotating arm principle are accounted for, and further gain variations caused by variations in track shape, depth and slope of the pits on the disc and varying quality of the transparent substrate protecting the disc.[3]

Finally, we consider variations in the undamped natural frequency of parasitic resonance modes (iv). From earlier robust control studies (6) it was learned that the resonances at 0.8, 1.6 and 4.3 kHz are especially important. To model these variations one could use additive or multiplicative frequency response embedding. However, this implies that the uncertainty is represented on an input-output basis rather than parametrically, which may result in an unnecessary design conservatism. In addition, high-order weighting functions may be required to arrive at an accurate bound on the frequency response. For this reason, we choose to model these frequency uncertainties as parametric perturbations in the nominal state-space model. To accomplish this we convert each second-order resonance mode in the A-matrix by a similarity transformation into the following real modal form[6]:

$$\begin{bmatrix} 0 & 1 \\ -\omega_0^2 & -2\beta\omega_0 \end{bmatrix}$$

The undamped natural frequency ω_0 of each mode appears as a parameter in the state-space model. The perturbation of the three resonance frequencies is carried out with the Parametric Uncertainty Modeling Toolbox.[7] The output of the Toolbox is a Linear Fractional Transformation description of the perturbed system

$$\dot{x} = Ax + B_1 u + B_2 w_p$$
$$y = C_1 x + D_{11}u + D_{12}w_p$$
$$z_p = C_2 x + D_{21}u + D_{22}w_p \tag{1}$$
$$w_p = \Delta_{\text{par}} z_p$$

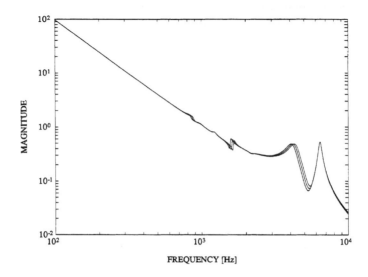

Figure 5. The nominal model and the parametric uncertainty set of $G_{11}(s)$.

with a normalized, block diagonal, parametric perturbation structure

$$\Delta_{\text{par}} = \text{diag}[\delta_1 I_1, \delta_2 I_2, \delta_3 I_3]$$

Each frequency perturbation involves a real-repeated perturbation block of multiplicity two. It is repeated because $w0$ appears quadratically in the A-matrix. The lower and upper bounds are chosen to be ±2.5% relative to their nominal values. In Figure 5, the resulting set of models for $G_{11}(s)$ is shown.

5.4. PERFORMANCE SPECIFICATIONS

A major disturbance source for the controlled system are track-position irregularities. Within the standardization of CDs, a radial track deviation of 100 μm (eccentricity) and a track acceleration at scanning velocity of 0.4 m/s^2 are allowed, while in the vertical direction these values are 1 mm and 10 m/s^2, respectively. Apart from track position irregularities, a second important disturbance source are external mechanical shocks. Measurements show that during portable use disturbance signals occur at a frequency band concentrated between 5 and 150 Hz.

For the CD player to work properly, the maximum allowable position error is 0.1 μm in the radial direction, and 1 μm in the focus direction. In the frequency domain, these performance specifications can be translated into requirements on the shape of the (output) sensitivity function $S = (I+GK)^{-1}$. Note that track irregularities involve time-domain constraints on signals which are difficult to translate into frequency domain specifications. To obtain the required track disturbance attenuation, the magnitude of the sensitivity at the rotational frequency should be less than 10^{-3} in both the radial and the focus direction. Further, for frequencies up to 150 Hz the sensitivity should be sufficiently small to suppress the impact of mechanical shocks and higher harmonics of the disc eccentricity.

Because both QFT and μ-synthesis are frequency domain techniques, the above specifications must be appropriately defined. In μ-synthesis, we require that the singular values of the sensitivity function be bounded from above by the values shown in Figure 6. In multivariable QFT, closed-loop specifications take on a different form. Specifically, we require that

$$|s_{ij}(j\omega)| \leq \eta_{ij}(\omega), \quad i, j = 1, 2, \quad \text{for each } G \in \mathcal{G}$$

as shown Figure 6. This implies that in multivariable QFT the degree of closed-loop interaction is explicitly defined.

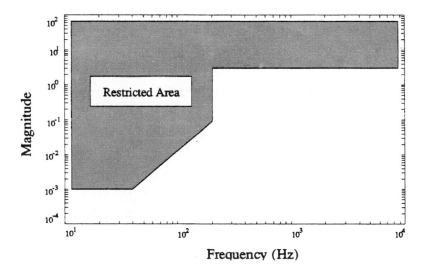

Figure 6. Sensitivity specifications.

Satisfaction of the above requirements at increasing rotational speeds requires a similar increase in the servo bandwidth. As stated earlier, a higher bandwidth has implications on the robustness of the design against manufacturing tolerances. In addition, higher bandwidth implies higher power consumption (very critical in portable use), generation of audible noise by the actuators and bad playability of discs with surface scratches. Therefore, besides a disturbance rejection objective we are always striving towards a lowest possible servo bandwidth (< 1 kHz).

In addition to these conflicting bandwidth requirements, the peak magnitude of the sensitivity function should be less than 3 to create sufficient phase margin in both loops. This is very important, since among several reasons the controller has to be discretized for implementation. Note that the invariance of the Bode sensitivity integral requires the trade-off between the sensitivity reduction and the phase margin specs.

5.5. μ-SYNTHESIS

5.6.1. Choice of Standard Plant

To begin with, the performance specification on S (sensitivity) can be combined with the robustness specifications associated with the transfer function KS accounting for the complex valued additive uncertainty Δa. The performance trade-offs are realized with a low-pass weighting function W1 on S, while imposing the weight W2 on KS reflects the size of the additive uncertainty and can also be used to force high roll-off at the input of the actuators. In this context, the objective to achieve robust performance implies[8] that for all stable, normalized perturbations Δ_a the closed loop is stable and

$$\| W_1 (I + [G + W_2 \Delta_a] K)^{1} \|_\infty < 1$$

This objective is exactly equal to the requirement that

$$\mu_\Delta (F_l(P_2, K)) < 1$$

with the interconnection

$$F_l(P_2, K) = \begin{bmatrix} W_2 K S & W_2 K S \\ W_1 S & W_1 S \end{bmatrix} \tag{2}$$

where $P2$ is the standard plant (including the plant model and the weightings). The corresponding uncertainty structure is $\Delta = \text{diag}[\Delta_a, \Delta_p]$, where Δ_p is a 2×2 performance block. Note that the related H_∞ problem $\| F_l(P_1, K) \|_\infty \le 1$ where

$$F_l(P_1, K) = \begin{bmatrix} W_2 K S \\ W_1 S \end{bmatrix} \tag{3}$$

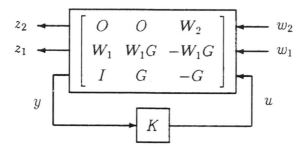

Figure 7. Standard plant for robust performance design.

a so-called "S, KS" design problem, yields robust *stability* and *only* nominal performance (although the original spec is robust performance). Standard H_∞ design on Equation (2), i.e. $\|F_l(P_2, K)\|_\infty \le 1$, introduces a factor $\sqrt{2}$ compared to Equation (3). Although it will achieve robust performance, it is clearly conservative since the known uncertainty structure is not accounted for. The μ-design for Equation (2) will yield robust performance as required.

An alternative that we will use is to specify performance on the transfer function SG instead of S, reducing the resulting controller order by 4^6:

$$F_l(P, K) = \begin{bmatrix} W_2 KS & W_2 KSG \\ W_1 S & W_1 SG \end{bmatrix} \qquad (4)$$

The standard plant is shown in Figure 7.

This design formulation has the additional advantage that it does not suffer from pole-zero cancellation as in the mixed sensitivity design formulations in Equation (2).

5.5.2. Application to the CD Mechanism

Using the DK iteration scheme[8,9], μ-controllers are synthesized for several design problems of increasing degree of difficulty.[10] Starting with the robust performance problem in Equation (4), the standard plant is augmented step by step with the parametric perturbations (ii), (iii) and (iv). In the final design problem we consider the complete uncertainty model, and the robust performance problem has 9 blocks:

$$\Delta = \mathrm{diag}[\delta_1 I_1, \delta_2 I_2, \delta_3 I_3, \delta_{i1}, \delta_{i2}, \delta_{o1}, \delta_{o2}, \Delta_a, \Delta_p]$$

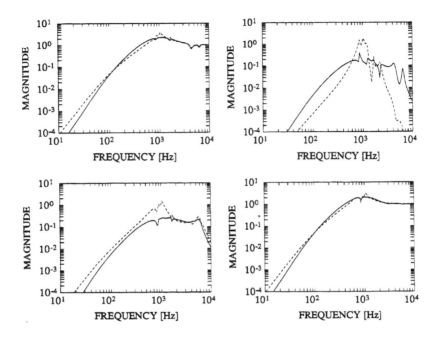

Figure 8. Nominal sensitivity with μ (- - -) and two PID controllers (- -).

In Figure 8, the sensitivity frequency response is shown for the full problem μ-controller. For comparison, the result is also shown for two (i.e. radial and focus) PID controllers achieving 800 Hz bandwidth in both loops. Clearly, the μ-controller achieves superior disturbance rejection up to 150 Hz, has lower interaction and a lower sensitivity peaking value.

The frequency responses of the controllers are shown in Figure 9. As expected, the μ-controller has higher low-frequency gain and actively acts upon the resonance frequencies.

5.5.3. Controller Reduction

The result of the μ-synthesis procedure for the full design problem is a controller with 83 states. This is primarily because of the dynamic D-scales necessary to account for the structure of the optimization problem. Of course, when the controller has to be experimentally validated this poses a problem. Digital implementation involves selection of the sampling frequency of the discretized controller.[11,12] Based on experience with previous implementations of SISO radial controllers (see[6]) and on the

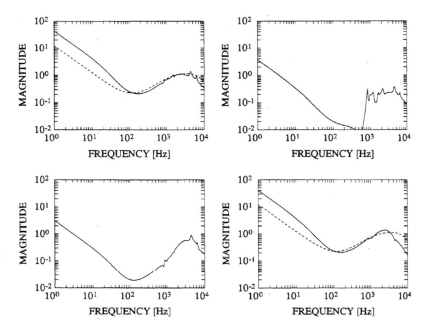

Figure 9. Frequency response of the μ-controller (- - -) and of two PID controllers (- - -).

location of the fastest poles in the multivariable μ-controller (8 kHz), a frequency of 40 kHz was selected. However, the DSP environment limited the order of the controller that could be implemented to 8 at this sampling frequency. Controllers with high order would lead to an overload of the DSP. This indicated a need for a dramatic controller order reduction, since there is a large gap between the practically allowable controller order (< 9) and the 83rd controller order computed using the μ-synthesis methodology. This posed a difficulty since the order of these controllers cannot be reduced to 8 without appreciable degradation of the nominal and robust performance. To facilitate implementation of higher-order controllers we can either design with a sampling frequency below 40 kHz, or use more than one DSP system and keep the sampling frequency at 40 kHz.

In this research we chose the latter, for it is expected that the necessary reduction of the sampling frequency sufficient for implementation would destroy the achievable performance. We have chosen to use two DSP boards as hardware set-up (each one with one input and two outputs, so as to realize all four parts of the MIMO controller). The controller to be implemented had to be order reduced from 83 to $2 \times 8 = 16$. The first step in the reduction procedure involved open-loop balancing followed by truncation of the (nearly) unobservable and/or uncontrollable state coordinates. The result of this balanced truncation was a 53rd order controller denoted $K_{53}(s)$. Inspection of the open-loop frequency response did not show any differences compared to the full-order controller. At all frequencies the maximum singular value of the additive reduction error was a factor 1000 or more smaller than the minimum singular value of the full-order controller.

The second step in the reduction procedure involved frequency weighted closed-loop balanced reduction of $K_{53}(s)$ as developed in[4,5]. We first briefly discuss below the frequency weighted closed-loop balanced reduction. Because the technique is completely dual for model and controller order reduction, we only discuss the controller reduction case.

The key idea is to use the same standard plant-controller interconnection structure used in H_∞ or μ-controller designs given by Equation (4), with identical weightings, but with the parametric perturbations left out (see Figure 7).

In a state-space realization of $F_l(P, K)$ the state vector can be divided into $x = [x_P; x_K]$ where x_P represents the state vector of the plant and the weightings and x_K represents the state of the controller. The design freedom offered in the choice of similarity transformation on K is exploited to find a closed-loop balanced $\mathbf{K}$ such that $\mathbf{P_K} = \mathbf{Q_K} = \text{diag}[\sigma_K]$ with $\mathbf{P_K}$ and $\mathbf{Q_K}$ being the lower right part (K-part) of the controllability Gramian and of the observability Gramian of $F_l(P, K)$ respectively. The Hankel Singular Values (HSV) σ_K are a measure for the contribution of the corresponding controller coordinates to the input-output behavior of the closed-loop configuration $F_l(P, K)$, and by truncating $\mathbf{K}$ only the most important controller dynamics in terms of closed-loop performance are retained. This method is an extension of Enns' frequency weighted model reduction for closed-loop systems, and is applied interactively (the Weighted Order Reduction Toolbox (Wortoolbox).[4] This enables appropriate reduction despite the absence of an a priori error bound.

This Toolbox was used to reduce $K_{53}(s)$ as follows:

STEP 1:   Write the open-loop controller $K_{53}(s)$ as two single-input two-output parts

$$K_{53}(s) = [K_{53}^1(s) \; K_{53}^2(s)] = C_K(sI - A_K)^{-1}[B_K^1 \; B_K^2] \tag{5}$$

in which $K_{53}^i(s), i = 1, 2$ is a 2×1 MTF (columns of $K_{53}(s)$).

STEP 2:   Construct the closed-loop configuration of Figure 10 in which the output vector is split into the radial and focus output. The standard plant $P(s)$ is fed back with both columns $K_{53}^1(s)$ and $K_{53}^2(s)$ of the controller.

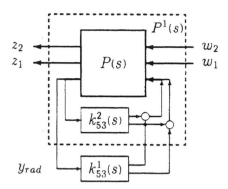

Figure 10. The closed-loop configuration used in the first step of frequency weighted closed-loop balanced reduction.

Regard $P(s)$ with $K^2_{53}(s)$ in the feedback loop as the new standard plant $P^1(s)$ that has to be controlled by $K^1_{53}(s)$. In this closed-loop setting, reduce the order of $K^1_{53}(s)$ using the frequency weighted closed-loop balanced reduction technique.

Since the $K^1_{53}(s)$-realization $C_K(sI - A_K)^{-1}B_K$ has smaller controllability measures than the $K_{53}(s)$-realization of Equation (5), there may be a possibility for a larger order reduction. Eventually, $K^1_{53}(s)$ should be reduced to $K^1_8(s)$ since it is the maximum order that can be implemented in a single DSP system operating at 40 kHz sampling frequency.

STEP 3: Once $K^1_{53}(s)$ has been reduced to $K^1_8(s)$, we can construct the dual closed-loop configuration with $P^2(s)$ fed back with $K^2_{53}(s)$ (see Figure 11), and reduce its order in this configuration. As before, $K^2_{53}(s)$ should be reduced to $K^2_8(s)$ for implementation on a single DSP system at a sampling frequency of 40 kHz.

Note that in $P^2(s)$, the CD-mechanism model is fed back with the already reduced-order $K^1_8(s)$ and not with the full-order $K^1_{53}(s)$. With this approach, the reduction result of the second column $K^2_{53}(s)$ accounts for the effects of the reduction of the first column. Note that this is not the case if we reduce $K^1_{53}(s)$ and $K^2_{53}(s)$ separately using open-loop reduction techniques.

An alternative to the above mentioned procedure is the application of the dual reduction procedure based on splitting the input vector u, i.e. row-wise separation of $K(s)$. However, we were not able to improve the result in this case.

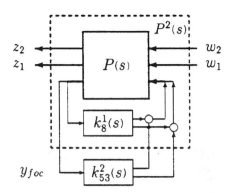

Figure 11. The closed-loop configuration used in the second step of frequency weighted closed-loop balanced reduction.

The open-loop frequency response of the full 83rd order μ-controller of the complete design and the 2×8 reduced order version K_{8+8} is shown in Figure 12. The difference between the two is largest above 2 kHz. Furthermore the (2,1) element is much different.

5.6.4. Practical Aspects of μ-Designs

The most significant experiences and limitations with respect to the use of μ-synthesis for this problem are:

- Convergence of the DK iteration scheme was fast (most cases 2 steps). Although global convergence of the scheme can not be guaranteed, in our design the resulting μ-controller did not depend on the starting controller.
- The final μ-controllers are of high order (83 for the full problem), due to the dynamic D-scales associated with the perturbations. This means that attention must be devoted to efficient order reduction of the final controller.
- Straightforward inclusion of the frequency perturbations into the standard plant caused numerical problems due to bad conditioning of the associated D-scales. For that reason, initially constant D-scaling was applied to Equation (1).
- The Toolbox used for control design[8] did not allow fitting of D-scales for repeated perturbations. Therefore, each repeated perturbation had to be replaced by two independent scalar perturbations. μ-analysis revealed that this did not affect the μ-bounds.

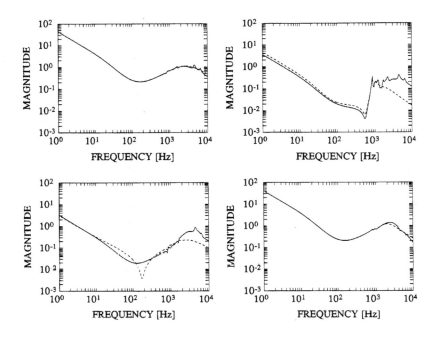

Figure 12. Open loop frequency response of $K_{83}(- - -)$ and $K_{8|8}$ (- -).

- Although most of the perturbations are real-valued by nature, the assumption in design that all perturbations are complex valued (and non-repeated) did not introduce much conservatism with respect to robust performance (see Figure 13). Note that the peak value of μ over frequency is 1.75, meaning that robust performance has not been achieved. This is due to the severe performance weighting W_1.

- The most difficult aspect of design appeared to be the shaping of weighting functions such that a proper trade-off is obtained between the conflicting performance and robustness specifications. In μ-synthesis the weighting functions are the "design knobs" available to the engineer. This also points out the importance of the modeling of disturbances and specifications.

- The frequency-range of interest (1 Hz to 10 kHz) in this application did cause numerical problems. The problem was resolved by applying internal balancing and time-scaling to the weighting functions and to the nominal model.

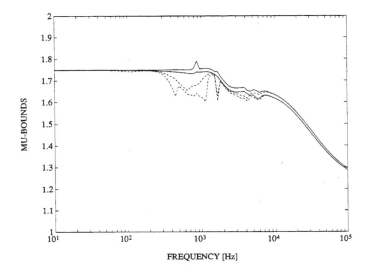

Figure 13. Complex (- - -) and real (- -) μ-bounds for the μ-controller on the standard plant including all perturbations.

5.6. QFT DESIGN

5.6.1. Problem Formulation

In multivariable QFT, the design problem is rigorously broken into an n-step sequential process ($n = 2$ in our design). At each sequential step, n SISO QFT problems are solved for the n closed-loop transfer functions in that loop. In this investigation, we have employed new multivariable QFT algorithms which do not involve plant inversion used in all other multivariable QFT techniques (e.g.,[13]). A brief overview of the new algorithms is given below (for complete details see[14]).

The sensitivity MTF is given by $S = (I + GK)^{-1} = \{s_{ij}\}, i, j = 1, ..., n$. The plant model G, an $n \times n$ MTF, is uncertain and belongs to the set G. The controller is diagonal, $K = \text{diag}[k1, ..., kn]$. The control objective is to design a controller, K, that in addition to achieving robust stability, satisfies the performance specification

$$|s_{ij}(j\omega)| \leq \eta_{ij}(\omega), \quad i, j = 1, 2, ..., n, \quad \text{for each } G \in \mathcal{G}$$

In the sequential procedure, at each ith step we use relations for $sij, j =$

1, ..., n, which are given in terms of the SISO controller ki[14]

$$s_{ij} = \frac{\alpha_{ji}^{L_1^{(i-1)}} + \sum_{k=i+1}^{n} -s_{kj}\alpha_{ki}^{L_1^{(i-1)}}}{\alpha_{ii}^{L_1^{(i-1)}} + \det[L_1^{(i-1)}]k_i}, \qquad i, j = 1, 2, \dots, n \qquad (6)$$

where $\alpha_{ij}^{L_1^{(i-1)}}$ is cofactor of $L_1^{(i-1)}$ (see[14]). Using the triangle inequality we can eliminate the unknowns s_{kj} (the closed-loop relations of loops not yet designed). The idea is to represent loop interactions as disturbances. The disturbances are represented in terms of closed-loop SISO elements of loopsyet to be designed. If we assume that there exists a solution to the design problem, then it possible to replace the unknown disturbance by its worst case magnitude.

The following are the standard n SISO QFT problems at the ith step

$$\left|s_{ij}\right| = \frac{\left|\alpha_{ji}^{L_1^{(i-1)}}\right| + \sum_{k=l+1}^{n} \eta_{kj}\left|\alpha_{ki}^{L_1^{(i-1)}}\right|}{\left|\alpha_{ii}^{L_1^{(i-1)}} + \det[L_1^{(i-1)}]k_i\right|}, \leq \eta_{ij},$$

$$i, j = 1, 2, \dots, n \quad \text{for each } G \in \mathcal{G} \qquad (7)$$

The robust stability criterion for the sequential, direct method requires that (under certain assumptions) each $1 + g_{ii}^i k_i$ is robust. Note that after the ith loop has been designed, the known controller k_i is sequentially embedded into the algorithms. Specifically, if $G = \{g_{ii}^i\}$, then at the ith loop design step the effective plant elements include all known controllers ($k_1, ..., k_{i-1}$) and are given by

$$g_{ij}^i = g_{ij}^1 - \sum_{k=1}^{i-1} \frac{g_{kj}^k g_{ik}^k k_k}{1 + g_{kk}^k k_k} \qquad (8)$$

To summarize the multivariable QFT technique, at each step in the sequential procedure we design a SISO controller to satisfy n SISO performance problems and one SISO stability problem.

5.6.2. Application to the CD Mechanism

In QFT we compute performance bounds only within a finite bandwidth (defined from the specifications), in this case, [0, 200] Hz. In addition, the constraint $|s_{ii}(\omega)| \leq 3, \omega \geq 0$, is used to generate stability margin constraints on each loop as follows

$$|1 + g_{ii}^i k_i| \geq \frac{1}{3}, \omega \geq 0, \quad \text{for each } G \in \mathcal{G} \qquad (9)$$

The final requirements are that the nominal open-loop bandwidth in each loop should be kept below 1 kHz to avoid amplifying high frequency noises and audible responses of the actuators, and that the controller order be kept to a minimum.

Note that although the open-loop system is weakly coupled, the specifications are truly multivariable since precise off-diagonal response is prescribed.

As discussed earlier, on the basis of experience and physical reasoning the nominal plant was assigned two types of uncertainties. The first type, parametric, consists of ±2.5% (independent) variations about nominal values in each of the undamped natural frequencies of the vibration modes at 0.8, 1.6 and 4.3 kHz. The second type, non-parametric, is related to the off-diagonal responses, and is defined as

$$G = \begin{bmatrix} g_{11} & \epsilon g_{22} + g_{12} \\ \epsilon g_{11} + g_{21} & g_{22} \end{bmatrix}, \epsilon \in [-0.1, 0.1]$$

Note that the g_{ij} terms above already include parametric uncertainties.

In multivariable QFT design, there are three questions that should be answered prior to an actual design:

1. Do we need to decouple the plant and/or reorder inputs/outputs?
2. What is the best sequential order of loop closure?
3. How to grid G to represent a parametric plant uncertainty model?

The open-loop plant is already weakly coupled and the order is already physically defined, hence the first question is naturally solved. Unfortunately, no general rules exists for solving the second question. However, in most cases, it is possible to define this sequence based on the relative degree of difficulty loosely defined from the plant frequency response and the specifications at each loop. In our problem, g_{11} and g_{22} have similar responses up to 1 kHz, and since they have identical stability and performance specifications at that range, we expect the corresponding QFT bounds to be near identical at that range. Between 1 and 10 kHz, however, the effect of the mechanical resonances is more pronounced in g_{11}, hence stability bounds for g_{11} will be more "demanding" than those for g_{22}. We concluded that the first loop should be closed first. For the final question, in our plant there are only 4 parameters, so a crude grid over the parameter space (three natural frequencies and the interaction factor) should be sufficient. We used 375 plant cases to represent the uncertainty.

The first design step involves the radial controller k_1 to satisfy robust stability and the robust sensitivity performance specifications. It is always recommended to first view the plant uncertainties in terms of templates, and those for

g_{11}^1 are shown in Figure 14 at $\omega = [62.8, 251.3, 1257, 3142, 5404, 10120,$ 26890 and 62830] rad/sec. Note that Figure 14 shows significant phase and magnitude variations at the three uncertain resonance frequencies.

To compute QFT stability bounds, we use the inequality defined by the margins problem in Equation (9)

$$\left| 1 + \frac{g_{11} + \det[G]k_2}{1 + g_{22}k_2} k_1 \right| \geq \frac{1}{3}$$

Because k_2 is yet to be designed, we consider its extreme values, namely,

$$\begin{cases} |1 + g_{11}k_1| \geq \dfrac{1}{3} \\[2mm] \left| 1 + \dfrac{\det[G]k_1}{g_{22}} \right| \geq \dfrac{1}{3} \end{cases}$$

To compute the sensitivity bounds invoke Equation (6) to obtain

$$\begin{cases} |S_{11}(j\omega)| = \dfrac{1 + \left| \dfrac{g_{12}s_{21}}{g_{22}} \right|}{\left| 1 + \dfrac{\det[G]k_1}{g_{22}} \right|} \leq \dfrac{1 + \left| \dfrac{g_{12}\eta_{21}}{g_{22}} \right|}{\left| 1 + \dfrac{\det[G]k_1}{g_{22}} \right|} \leq \eta_{11}(\omega) \\[6mm] |S_{12}(j\omega)| = \dfrac{\left| -\dfrac{g_{12}}{g_{22}} \right| \left| \dfrac{g_{12}s_{22}}{g_{22}} \right|}{\left| 1 + \dfrac{\det[G]k_1}{g_{22}} \right|} \leq \dfrac{\left| \dfrac{k_{12}}{k_{22}} \right| + \left| \dfrac{g_{12}\eta_{22}}{g_{22}} \right|}{\left| 1 + \dfrac{\det[G]k_1}{g_{22}} \right|} \leq \eta_{12}(\omega) \end{cases}$$

Robust stability bounds and robust sensitivity bounds were computed using slightly modified functions from the QFT Toolbox for MATLAB.[15] These bounds and the synthesized nominal loop are shown on the Nichols chart in Figure 15.

Note that $L_{11,0}(s) = g_{11,0}(s)k_1(s)$ where $g_{11,0}(s)$ is the nominal plant. The controller is given by (rounded off)

$$k_1 = \frac{1.33 \times 10^{10}(s^2 + 2.8 \times 10^3 s + 4.6 \times 10^6)(s + 12.1)}{(s + 2.9 \times 10^5)(s^2 + 2.6 \times 10^4 s + 3.8 \times 10^8)(s + 96)(s + 1)}$$

The second design step is to design k_2 to satisfy the robust stability and the robust performance specification while taking into account the known k_1. The plant used in the second design step, g_{22}^2, is defined by Equation (8)

$$g_{22}^2 = g_{22} - \frac{g_{12}g_{21}k_1}{1 + g_{11}k_1}$$

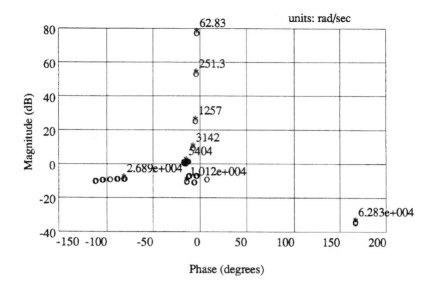

Figure 14. Templates of g_{11}^1 used in the first step.

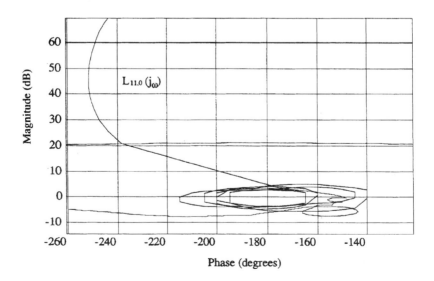

Figure 15. QFT bounds and nominal design $L_{11,0}(s)$ in the first design step.

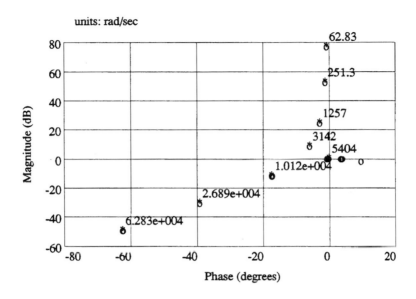

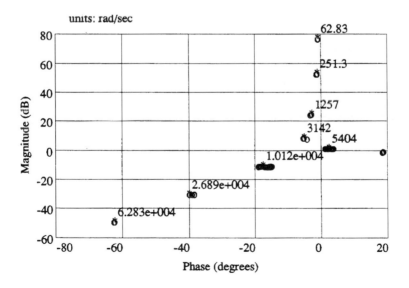

Figure 16. Templates of g_{22} (top) and g_{22}^2 (bottom) used in the second step.

and its templates are shown in Figure 16. Note that due to the interaction, the templates of g_{22}^2 exhibit larger variations around the three uncertain natural frequencies compared with the templates g_{22} alone. Plant g_{22}^2 indeed is the plant seen by the controller to be designed.

Robust stability bounds are computed from Equation (9)

$$\begin{cases} \left| 1 + \dfrac{g_{22} + \det[G]k_1}{1 + g_{11}k_1}k_2 \right| \geq \dfrac{1}{3} \\[4mm] \left| 1 + \dfrac{g_{11} + \det[G]k_2}{1 + g_{22}k_1}k_2 \right| \geq \dfrac{1}{3} \end{cases}$$

At this point, the only unknown is g_2. Sensitivity bounds are computed from Equation (8)

$$\begin{cases} |s_{21})j\omega)| = \left| \dfrac{\dfrac{-g_{21}k_1}{1 + g_{11}k_1}}{1 + \dfrac{g_{22} + \det[G]k_1}{1 + g_{11}k_1}k_2} \right| \leq \eta_{21}(\omega) \\[8mm] |s_{22}(j\omega)| \leq \left| \dfrac{1}{1 + \dfrac{g_{22} + \det[G]k_1}{1 + g_{11}k_1}k_2} \right| \leq \eta_{22}(\omega) \end{cases}$$

The robust stability bounds, robust sensitivity bounds and the synthesized nominal loop are shown on the Nichols chart in Figure 17.

Note that the nominal loop in the second design step is $L_{22,0}(s) = g_{22}^2 k_1(s)$ where

$$g_{22,0}^2 = g_{22}^2 - \frac{g_{12,0}g_{21,0}k_1}{1 + g_{11,0}k_1}$$

The controller is given by (rounded off)

$$k_2 = \frac{2.9 \times 10^4(s + 2713.2)(s + 1622.2)}{s^2 + 6.4 \times 10^3 s + 2.9 \times 10^8)(s + 1)}$$

As expected, the robust stability bounds in the second loop are less "demanding" compared with those obtained in the first loop. Specifically, in the crossover frequency range, the first nominal loop is required to maintain less phase lag compared with the demand on the second nominal loop.

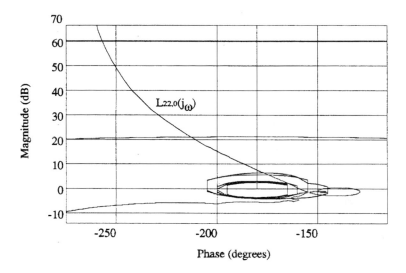

Figure 17. QFT bounds and nominal design $L_{22,0}(s)$ in the second step.

5.6.3. Practical Aspects of Multivariable QFT Designs

The most significant experiences and limitations with respect to the use of multivariable QFT for this problem are as follows.

- Parametric uncertain plant models, such as the model used for our CD mechanism, must be converted into plant templates. Forming a grid in the parameter space with less than 100 points is a reasonable approach for plants with at most 2–3 uncertain parameters. However, the variations in the natural frequencies of the lightly damped modes require use of a finer grid. However, too many template points (say > 1000) will slow down numerical computation of QFT bounds. One alternative, used here, was to study the relation between the grid and the resulting templates. The goal is to remove any points in the parameter space that correspond to interior plant template points. It was found out that a grid of 5 cases for each natural frequency and a grid of 3 for the off-diagonal gain is a minimum grid. This relatively course grid required careful consideration of the resulting bounds during loop shaping (i.e., avoidance of bounds).
- Manual loop shaping of k_1 was not a trivial task due to the location of nonminimum phase zero relative to the servo bandwidth. It required familiarity with conditionally stable loops and the concept of "phase lag maximization". It was found that controllers designed by other techniques

(e.g., μ) can be effectively used to both study their general response as an initial design or fine-tuning within the QFT environment.

- It is yet to be established that an off-diagonal controller can result in an improved design in our problem, hence, use of a diagonal controller may be justified. However, the QFT technique does not "provide" general instructions to the designer as to how to exploit the inherent interaction in the plant in order to improve closed-loop performance (our present research focuses on this issue).

5.7. IMPLEMENTATION RESULTS

The actual implementation of the controllers in the DSP systems has been carried out with the dSPACE Cit-Pro software.[11] Most problems occurring when implementing controllers involve scaling inputs and outputs to obtain the appropriate signal levels and to ensure that the resolution of the DSP is optimally used. These problems can be solved in a user-friendly manner using the dSPACE software. The two DSP systems used were: a TMS320C30 16 bit processor with floating point arithmetic and a TMS320C25 16 bit processor with fixed point arithmetic. The Analog Digital Converters (ADC) are also 16 bit and have a conversion time of 5 μs. The maximum input voltage is ± 10 V. The Digital Analog Converters (DAC) are 12 bit, have a conversion time of 3 μs and also operate within the range of ± 10 V.

The first column $K_8^1(s) = [K_{11}\ K_{12}]^T$ has been implemented in the TMS320C30 processor at a sampling frequency of 40 kHz. The dSPACE software provides for:

- A ramp-invariant discretization,
- A modal transformation, and
- Generation of C code that can be downloaded to the DSP.

The second column $K_8^2(s) = [K_{21}\ K_{22}]^T$ was implemented on the TMS320C25 processor. Since this processor has fixed point arithmetic, scaling is more involved.[12] With the dSPACE software the following steps have been carried out:

- A ramp-invariant discretization,
- A modal transformation,
- l_1 scaling for the non-integrator states,
- State scaling for the integrator states such that their contribution is at most 20%, and
- Generation of DSPL code that can be downloaded to the DSP.

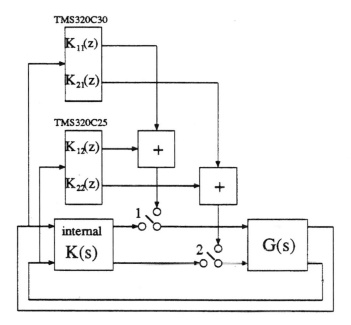

Figure 18. The connection of both DSP systems to the experimental set up.

For the TMS320C25, a discretization at a sampling frequency of 40 kHz appeared to be too high, since it resulted in a processor load of 115%. For that reason, the sampling frequency of this DSP has been lowered to 34 kHz, yielding a processor load of 98.6%.

The DSP systems have been connected to the experimental set up by means of two analog summing junctions that have especially been designed for this purpose (Figure 18).

When the external 2×8 μ-controller of the complete design is connected to the experimental set up, we can measure the achieved performance in terms of the frequency response of the sensitivity function. The measurements have been carried out using a Hewlett Packard 3562 Dynamic Signal Analyzer. Since this analyzer can measure only SISO frequency responses, each of the four elements of the sensitivity function must be determined separately.

The measurements are started each time at the same radial position on the disc. This position is chosen approximately half way the disc, since the model used for control design was identified at that location, and also the radial gain is most constant in this region. In Figure 19 the measured and simulated frequency response of the sensitivity function are shown. The off-diagonal

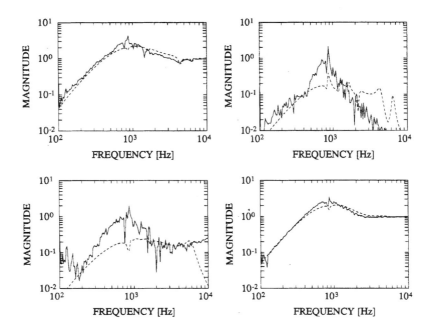

Figure 19. Measured frequency response of the sensitivity function with μ-controller K_{8+8} (---) and simulated with K_{83} (- - -).

elements are not very reliable since the coherence during these measurements was very low.

The nominal performance has been tested in terms of the possibility to increase the rotational frequency of the Compact Disc. It appeared possible to achieve an increase in speed of a factor 4.

The diagonal multivariable QFT controller was implemented using two DSP boards as done with the μ-controller. The discretization was done at a sampling rate of 27 kHz. The measured closed-loop sensitivity magnitudes are shown in Figure 20.

The observed peaks in $s_{11}(j\omega)$, especially in the 800–2000 Hz range, is attributed to identification errors in the nominal plant. A detailed comparison between the identified radial loop and the measured data around those peaks (Figure 3) revealed that although the magnitudes matched well the phases were off by a few degrees. A SISO QFT design was subsequently carried out and implemented in the radial loop using the raw experimental frequency response data (since with QFT the designer can work directly with measured frequency response data). The new design did not exhibit such peaks.[16]

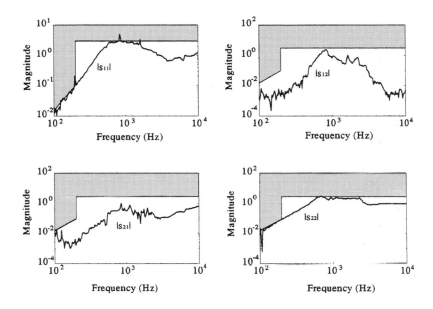

Figure 20. Measured frequency response magnitudes of the sensitivity functions with QFT controller.

5.8. CONCLUSIONS

In this paper multivariable QFT and μ-synthesis techniques have been applied to a Compact Disc player. The design objective was to achieve good track-following and focusing performance in the presence of disturbances and structured uncertainty in the lightly damped plant dynamics. Experimental results show that with use of the two design techniques, considerable improvement can be achieved in the track-following quality of disc drives. Such an improvement allows for a significant increase in the rotational frequency of the disc.

References

1. Bouwhuis, G. *et al.*, 1985, *Principles of Optical Disc Systems*, Adam Hilger Ltd., Bristol, UK.
2. Draijer, W.M., M. Steinbuch and O.H. Bosgra, 1992, *IFAC Automatica*, **28**, 455–462.
3. Schrama, R., 1992, *Approximate Identification and Control Design*, PhD Thesis, Delft Univ. Technology, The Netherlands.
4. Wortelboer, P., 1994, *Frequency Weighted Balanced Reduction of Closed-loop Mechanical Servo-systems: Theory and Tools*, PhD Thesis, Delft Univ. Technology, The Netherlands.

5. Wortelboer, P. and O.H. Bosgra, 1992, *Procs. IEEE CDC*, 2848–2849.
6. Steinbuch, M., G. Schootstra and O.H. Bosgra, 1992, *Procs. IEEE CDC*, 2596–2600.
7. Lambrechts, P., J.C. Terlouw, S. Bennani and M. Steinbuch, 1993, *Procs. ACC*, 267–272.
8. Balas, G.J., J.C. Doyle, K. Glover, A.K. Packard and R. Smith, 1991, *μ-Analysis and Synthesis Toolbox*, MUSYN Inc., Minneapolis.
9. Doyle, J.C., 1984, *Advances in Multivariable Control*, Lecture Notes of the ONR/Honeywell Workshop, Honeywell, Minnesota.
10. Groos, P.J.M., M. Steinbuch and O.H. Bosgra, 1993, *Procs. 2nd European Control Conference*, 981–985.
11. dSPACE GmbH, 1989, *DSPCitPro software package*, Documentation of dSPACE, GmbH, Germany.
12. Steinbuch, M., G. Schootstra, and H.T. Goh, 1994, *IEEE Trans. on Control Systems Technology*, 2, 312–317.
13. Horowitz, I., 1991, *Int. J. Control*, 53, 255–291.
14. Park M.S., 1994, *A New Approach to Multivariable Quantitative Feedback Theory*,ÿ20 Ph.D. Thesis, Univ. of Massachusetts, MA.
15. Borghesani, C., Y. Chait, and O. Yaniv, 1995, *The QFT Toolbox for MATLAB*, The MathWorks, Inc., USA.
16. Chait, Y., M.S. Park, and M. Steinbuch, 1994, *J. Systems Engineering*, 4, 107–117

6 A COMPONENT MODES DAMPING ASSIGNMENT METHODOLOGY FOR ARTICULATED MULTI-FLEXIBLE BODY SYSTEMS

ALLAN Y. LEE

Jet Propulsion Laboratory, California Institute of Technology, Pasadena, California 91109-8099, USA

6.1. MULTI-FLEXIBLE BODY SYSTEMS

Multi-flexible body systems are assemblages of rigid and flexible bodies where each body can rotate and/or translate relative to one another. Examples are spacecraft, robotic manipulators, helicopters, and other systems. The equations of motion of these systems are typically very complex. As a result, a multi-body simulation package such as DISCOS[1] must be used to determine the time responses of these systems when they are subjected to external dynamic stimuli. To this end, one must supply appropriate models for both the rigid-body and the flexible components involved. For complex systems such as the Galileo spacecraft, practical considerations (e.g., simulation time) impose limits on the number of modes that each flexible body can retain in a given simulation. Reduced-order models for these flexible components are hence needed.

The enhanced projection and assembly[2] and the component modes projection and assembly[3] model reduction methodologies are two effective ways of generating reduced-order component models for articulated multi-flexible body systems. The component modes projection and assembly model reduction methodology (COMPARE)[3] can generate reduced-order component models which when reassembled lead to a reduced-order system model which captured exactly dominant system modes as well as the static gain of a given input-output pair in the full-order system model. This methodology will be reviewed shortly.

In addition to the reduced-order component models, damping matrices of the flexible components are also needed as inputs to DISCOS. The treatment of damping in complex structures has always posed difficulties. Owing to the variety of mechanisms that contributed to damping, it has not been possible to model damping on a finite-element basis the way mass and stiffness are modeled. For simplicity, it is a common practice to introduce damping only after equations of motion (for all flexible components) have been transformed to their "modal" equivalences. Then, diagonal modal damping matrices with empirically determined damping factors are generated and added to these modal equations of motion. These component-level damping matrices are then combined to generate the system damping matrix.[4-5]

Experience obtained from working with today's space structures indicated that a uniform damping factor of about 0.25-0.50% can be conservatively assumed for all the system modes.[6-7] However, the system damping matrix obtained by combining component damping matrices with empirically determined damping factors might not adhere to this criterion. If this is the case, a time-consuming iterative procedure must then be used to adjust the damping factors of numerous component modes until the damping factors of the system modes do meet the criterion. Reference 7 described one such iterative procedure. However, that technique cannot be easily applied on large-order complex structures. A more practical and systematic approach is proposed here.

6.2. THE COMPARE METHODOLOGY REVISITED

Before the component modes damping assignment technique is described, let us first review the COMPARE methodology.[3] This is a two-stage model reduction methodology, combining the classical Component Mode Synthesis (CMS) method and the newly developed Enhanced Projection and Assembly method (EP&A).[2] A graphical illustration of the steps involved in this methodology is depicted in Figure 1.

In the first stage of the COMPARE methodology, CMS mode sets (such as the MacNeal-Rubin or Craig-Bampton mode sets) are generated for all the flexible components of the multi-flexible body system. Methods commonly used in generating these mode sets are well documented, and are not given here.[8] These mode sets are then used to reduce the full-order component

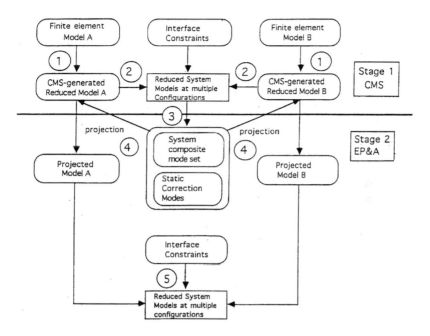

Figure 1. Graphical illustration of the COMPARE stages.

models. Once the reduced-order component models are generated, they are assembled using the interface displacement compatibility relations to produce reduced-order system models at multiple system configurations (see Step 2 in Figure 1).

To illustrate Step 2, consider a system with two flexible components. The undamped motion of component A is described by the following equations generated using the CMS method:

$$I_{pp}\ddot{\eta}_p^A + \Lambda_{pp}^A \eta_p^A = G_{pa}^A u_a^A \,,$$
$$y_b^A = H_{bp}^A \eta_p^A \,. \tag{1}$$

Here η_p^A and Λ_{pp}^A are the modal coordinate vector and a diagonal stiffness matrix of component A, respectively. Note that the dimensions of the matrix are indicated by its subscripts. For example, the matrix G_{pa}^A is a $p \times a$ control distribution matrix, and u_a^A is a $a \times 1$ control vector. Similarly, the matrix H_{bp}^A is a $b \times p$ output distribution matrix and y_b^A is an output vector. Similar equations can be written for the component B model:

$$I_{qq}\ddot{\eta}_q^B + \Lambda_{qq}^B \eta_q^B = G_{qa}^B u_a^B ,$$
$$y_l^B = H_{lq}^B \eta_q^B , \tag{2}$$

where q is the degree-of-freedom of the component B model.

The system equations of motion are next constructed using these component equations, and by enforcing displacement compatibility at the system interface. To this end, let $P \triangleq [P_{pe}^{A^T}\ P_{qe}^{B^T}]^T$ be a full-rank matrix mapping a minimal system state η_e into:

$$\begin{bmatrix} \eta_p^A \\ \eta_q^B \end{bmatrix} = \begin{bmatrix} P_{pe}^A \\ P_{qe}^B \end{bmatrix} [\eta_e] , \tag{3}$$

where η_e is a $e \times 1$ system coordinate vector ($e = p + q - i$), and i is the number of interface constraints. One way to generate the P matrix will be given in equations (11)-(14). Substituting $\eta_p^A = P_{pe}^A \eta_e$ and $\eta_q^B = P_{qe}^B \eta_e$ into (1) and (2), respectively, pre-multiplying the resultant equations by $P_{pe}^{A^T}$ and $P_{qe}^{B^T}$, and summing the resultant equations give:

$$M_{ee}\ddot{\eta}_e + K_{ee}\eta_e = G_{ea}u_a , \tag{4}$$
$$y_s = H_{se}\eta_e . \tag{5}$$

In (4-5), $M_{ee} = P_{pe}^{A^T} P_{pe}^A + P_{qe}^{B^T} P_{qe}^B$, $K_{ee} = P_{pe}^{A^T} \Lambda_{pp}^A P_{pe}^A + P_{qe}^{B^T} \Lambda_{qq}^B P_{qe}^B$, and $G_{ea} = P_{pe}^{A^T} G_{pa}^A + P_{qe}^{B^T} G_{qa}^B$. The system's output vector is $y_s = [y_b^{A^T}\ y_l^{B^T}]^T$, where $s = b + l$. The output distribution matrix is $H_{se} = [P_{pe}^{A^T} H_{bp}^A\ P_{qe}^{B^T} H_{lq}^{B^T}]^T$. To arrive at the expression for G_{ea}, we have assumed that $u_a^A = u_a^B = u_a$. Otherwise, the term $G_{ea} u_a$ in (4) should be replaced by $[P_{pe}^{A^T} G_{pa}^A\ P_{qe}^{B^T} G_{qa}^B][u_a^{A^T}\ u_a^{B^T}]^T$.

Let $(\Phi_{ee}, \Lambda_{ee}) = \text{eig}(K_{ee}, M_{ee})$, the "modal" equivalence of (4-5) is

$$I_{ee}\ddot{\phi}_e + \Lambda_{ee}\phi_e = \{\Phi_{ee}^T G_{ea}\}u_a ,$$
$$y_s = \{H_{se}\Phi_{ee}\}\phi_e , \tag{6}$$

where ϕ_e is a modal coordinate vector ($\eta_e = \Phi_{ee}\phi_e$), and $\Lambda_{ee} = \Phi_{ee}^T K_{ee}\Phi_{ee}$ is a diagonal matrix with the undamped system eigenvalues on its diagonal.

Note that the reduced-order system model given by (6) was obtained without using any system-level input-to-output mapping information. This drawback is remedied in the second stage of COMPARE using the EP&A methodology which does make use of the known system-level mapping information.

In using the EP&A methodology, only k of the system's e modes are kept while the remaining t (= $e - k$) modes are truncated (see Step 3 in Figure 1). The kept mode set is a "composite" mode set, consisting of "dominant" system modes found at all system configurations of interest, and not just at one particular configuration. This "composite" mode set approach had been proven to be effective in capturing important modes at all system configurations of interest.[2] With this understanding, we have:

$$\eta_e = \Phi_{ee}\phi_e \triangleq [\; \Phi_{ek} \quad \Phi_{et}\;] \begin{bmatrix} \phi_k \\ \phi_t \end{bmatrix} \doteq \Phi_{ek}\phi_k \, , \tag{7}$$

where ϕ_k and ϕ_t are generalized coordinate vectors associated with the "kept" and "truncated" modes, respectively, and Φ_{ek} and Φ_{et} are the corresponding eigenvector matrices.

Before the composite mode set Φ_{ek} is projected onto the component models generated in the first stage, it is first augmented with static correction modes Φ_{ea}. One way to generate these static correction modes was described in[2]. While the dimension of the augmented composite mode set is increased slightly by these static correction modes, they help to ensure that the static gain of the full-order system, for a given input-output pair, is exactly preserved in the reduced-order system model.

The projections of the augmented composite mode set $\Phi_{em} \triangleq [\Phi_{ek} \; \Phi_{ea}]$ ($m = k + a$) onto the CMS-generated component models are accomplished as follow (see Step 4 in Figure 1):

$$\begin{aligned} \eta_p^A &\doteq \{P_{pe}^A \, \Phi_{em}\} \phi_m^A \triangleq \Psi_{pm}^A \phi_m^A \, , \\ \eta_q^B &\doteq \{P_{qe}^B \, \Phi_{em}\} \phi_m^B \triangleq \Psi_{qm}^B \phi_m^B \, , \end{aligned} \tag{8}$$

where ϕ_m^A and ϕ_m^B denote generalized coordinate vectors for components A and B, respectively. Substitutions of (8) into (1) and (2), premultiplying the resultant equations by Ψ_{pm}^A and Ψ_{qm}^B, respectively, produce the reduced-order equations for these components:

$$\begin{aligned} \{\Psi_{pm}^{A^T}\Psi_{pm}^A\} \, \ddot{\phi}_m^A + \{\Psi_{pm}^{A^T}\Lambda_{pp}^A\Psi_{pm}^A\} \, \phi_m^A &= \{\Psi_{pm}^{A^T}G_{pa}^A\} \, u_a \, , \\ \{\Psi_{qm}^{B^T}\Psi_{qm}^B\} \, \ddot{\phi}_m^B + \{\Psi_{qm}^{B^T}\Lambda_{qq}^B\Psi_{qm}^B\} \, \phi_m^B &= \{\Psi_{qm}^{B^T}G_{qa}^B\} \, u_a \, . \end{aligned} \tag{9}$$

In general, neither the mass nor stiffness matrices in (9) is diagonal. To diagonalize these matrices, two eigenvalue problems associated with (9) are solved. Let Ξ_{mm}^A and Ξ_{mm}^B be the mass-normalized eigenvector matrices obtained, and v_m^A and v_m^B be the corresponding modal coordinate vectors. The modal equivalences of (9) are:

$$\begin{aligned} I_{mm}^A \ddot{v}_m^A + C_{mm}^A \dot{v}_m^A + \Lambda_{mm}^A v_m^A &= \{\Pi_{pm}^{A^T} G_{pa}^A\} \, u_a \, , \\ I_{mm}^B \ddot{v}_m^B + C_{mm}^B \dot{v}_m^B + \Lambda_{mm}^B v_m^B &= \{\Pi_{qm}^{B^T} G_{qa}^B\} \, u_a \, . \end{aligned} \tag{10}$$

In (10), $\Pi_{pm}^A \overset{\Delta}{=} \Psi_{pm}^A \Xi_{mm}^A$, and $\Lambda_{mm}^A = \Pi_{pm}^{A^T} \Lambda_{pp}^A \Pi_{pm}^A$ is a diagonal eigenvalue matrix of the reduced-order component A model. The matrix Λ_{mm}^B is that for component B model. Modal damping terms, $C_{mm}^A \dot{v}_m^A$ and $C_{mm}^B \dot{v}_m^B$, had also been "added" in (10). These damping terms are simplified representations of several energy dissipation mechanisms such as Coulomb friction, hysteresis damping, etc. The hypothesis being that, for lightly damped structures, these energy dissipation mechanisms can be grossly represented by the normal damping matrices C_{mm}^A and C_{mm}^B in (10). However, the component modes damping assignment methodology to be described next is equally applicable to systems where the component damping matrices C_{mm}^A and C_{mm}^B are not "normal."

In the last step (see Step 5 in Figure 1), reduced-order system equations of motion are generated by enforcing displacement compatibility at the system interface:

$$N_{ip}^A \eta_p^A + N_{iq}^B \eta_q^B = O_i , \tag{11}$$

where N_{ip}^A and N_{iq}^B are matrices that established the displacement compatibility relations between components A and B. These equations can be re-written in terms of the modal coordinate vectors v_m^A and v_m^B:

$$[N_{ip}^A \Pi_{pm}^A \quad N_{iq}^B \Pi_{qm}^B] \begin{bmatrix} v_m^A \\ v_m^B \end{bmatrix} \overset{\Delta}{=} [D_{ic}] \begin{bmatrix} v_m^A \\ v_m^B \end{bmatrix} \doteq O_i . \tag{12}$$

The compatibility matrix D_{ic} ($c = 2m$) can be partitioned using the Singular Value Decomposition (SVD) technique:

$$D_{ic} = [U_{ii}] [\Sigma_{ii} , O_{id}] \begin{bmatrix} P_{ci}^T \\ P_{cd}^T \end{bmatrix} , \tag{13}$$

where $d = c - i$, Σ_{ii} is an $i \times i$ diagonal matrix with the i singular values of the matrix D_{ic} along its diagonal, and O_{id} is an $i \times d$ null matrix. The partitioned matrix P_{cd} in (13) can be used as follows:[9]

$$\begin{bmatrix} v_m^A \\ v_m^B \end{bmatrix} = P_{cd} v_d \overset{\Delta}{=} \begin{bmatrix} P_{md}^A \\ P_{md}^B \end{bmatrix} v_d . \tag{14}$$

Here, v_d denotes a minimum set of generalized coordinates of the reduced-order system model. To obtain the reduced-order system equations of motion, we substitute $v_m^A = P_{md}^A v_d$ and $v_m^B = P_{md}^B v_d$ into (10), pre-multiplying the resultant equations by $P_{md}^{A^T}$ and $P_{md}^{B^T}$, respectively, and summing the results:

$$M_{dd} \ddot{v}_d + \tilde{C}_{dd} \dot{v}_d + K_{dd} v_d = G_{da} u_a ,$$
$$y_s = H_{sd} v_d , \tag{15}$$

where $M_{dd} = P_{md}^{A^T} P_{md}^A + P_{md}^{B^T} P_{md}^B$, $\tilde{C}_{dd} = P_{md}^{A^T} C_{mm}^A P_{md}^A + P_{md}^{B^T} C_{mm}^B P_{md}^B$, and similar expressions could also be written for the matrices K_{dd}, G_{da}, and H_{sd}. Let $(\Phi_{dd}, \Lambda_{dd}) = \text{eig}(K_{dd}, M_{dd})$, the modal equivalence of (15) is:

$$I_{dd}\ddot{\chi}_d + C_{dd}\dot{\chi}_d + \Lambda_{dd}\chi_d = \{\Phi_{dd}^T G_{da}\}u_a, \qquad (16)$$

$$y_s = \{H_{sd}\Phi_{dd}\}\chi_d, \qquad (17)$$

where χ_d is a modal coordinate vector ($v_d = \Phi_{dd}\chi_d$), $C_{dd} = \Phi_{dd}^T \tilde{C}_{dd}\Phi_{dd}$, and $\Lambda_{dd} = \Phi_{dd}^T K_{dd}\Phi_{dd}$ is a diagonal matrix with the reduced-order system eigenvalues along its diagonal. It was proven in Ref. 3 that Φ_{ek} (in (7)) is captured exactly in Φ_{dd} in spite of augmenting the composite mode set with static correction modes. The effectiveness of the COMPARE methodology had been successfully verified using a high-order finite-element model of the Galileo spacecraft.[3]

Experience obtained from working with today's space structures indicated that a uniform damping factor of about 0.25–0.50% can usually be conservatively assumed for all the system modes. The system damping matrix C_{dd} obtained by combining component damping matrices C_{mm}^A and C_{mm}^B is unlikely to adhere to this criterion. A way to determine the component damping factors that will lead to a system damping matrix with the desired properties is hence needed. The component modes damping assignment methodology serves this need.

6.3. COMPONENT MODES DAMPING ASSIGNMENT METHODOLOGY

To address the problem we have just posed, a relation between the component and system damping matrices must first be established. To this end, we note from (14) that $v_m^A = P_{md}^A v_d = \{P_{md}^A \Phi_{dd}\}\chi_d \triangleq Q_{md}^A \chi_d$ and $v_m^B = \{P_{md}^B \Phi_{dd}\}\chi_d \triangleq Q_{md}^B \chi_d$. Substituting these relations into (10), pre-multiplying the resultant equations by $Q_{md}^{A^T}$ and $Q_{md}^{B^T}$, respectively, equating the sum of the resultant equations to (16), and comparing like terms give:

$$Q_{md}^{A^T} Q_{md}^A + Q_{md}^{B^T} Q_{md}^B = I_{dd}, \qquad (18)$$

$$Q_{md}^{A^T} C_{mm}^A Q_{md}^A + Q_{md}^{B^T} C_{mm}^B Q_{md}^B = C_{dd}, \qquad (19)$$

$$Q_{md}^{A^T} \Lambda_{mm}^A Q_{md}^A + Q_{md}^{B^T} \Lambda_{mm}^B Q_{md}^B = \Lambda_{dd}. \qquad (20)$$

More insight is gained if these equations are re-written as follow. Denoting:

$$Q_{md}^{A^T} \triangleq [q_{d1}^{A_1} \quad q_{d1}^{A_2} \quad \cdots \quad q_{d1}^{A_m}], \qquad (21)$$

$$Q_{md}^{B^T} \triangleq [q_{d1}^{B_1} \quad q_{d1}^{B_2} \quad \cdots \quad q_{d1}^{B_m}], \qquad (22)$$

where $q_{d1}^{A_i}$ and $q_{d1}^{B_j}$, $(i,j) = 1,...,m$ are $d \times 1$ vectors. Defining:

$$R_{dd}^{A_i} \triangleq q_{d1}^{A_i} q_{d1}^{A_i^T} , \qquad (23)$$

$$R_{dd}^{B_j} \triangleq q_{d1}^{B_j} q_{d1}^{B_j^T} , \qquad (24)$$

where $R_{dd}^{A_i}$ and $R_{dd}^{B_j}$, $(i,j) = 1,...,m$ are $d \times d$ rank one symmetric matrices. Using (21-24), the matrix algebraic relations in (18-20) become:[10]

$$\sum_{i=1}^{i=m} R_{dd}^{A_i} + \sum_{j=1}^{j=m} R_{dd}^{B_j} = I_{dd} , \qquad (25)$$

$$\sum_{i=1}^{i=m} c^{A_i} R_{dd}^{A_i} + \sum_{j=1}^{j=m} c^{B_j} R_{dd}^{B_j} = C_{dd} , \qquad (26)$$

$$\sum_{i=1}^{i=m} \lambda^{A_i} R_{dd}^{A_i} + \sum_{j=1}^{j=m} \lambda^{B_j} R_{dd}^{B_j} = \Lambda_{dd} . \qquad (27)$$

To arrive at (26), we have assumed that the component damping matrices are diagonal. That is, $C_{mm}^A = \text{diag}[c^{A_1}, ..., c^{A_m}]$ and $C_{mm}^B = \text{diag}[c^{B_1}, ..., c^{B_m}]$. This is a simplifying assumption. If, for example, the (1,2) and (2,1) elements of C_{mm}^A (denoted by $c^{A_{12}}$) are non-zero, then, a term $c^{A_{12}} (q_{d1}^{A_1} q_{d1}^{A_2^T} + q_{d1}^{A_2} q_{d1}^{A_1^T})$ must be added to the left-hand-side of (26). In (27), λ^{A_i} and λ^{B_j} are the squared frequencies of the component modes.

The matrices $R_{dd}^{A_i}$ and $R_{dd}^{B_j}$ in (26-27) could be interpreted as "relative contribution" matrices. The matrix $R_{dd}^{A_i}$ determines the contribution made by the damping factor of the component A's i^{th} mode in the system damping matrix. Similarly, the matrix $R_{dd}^{B_j}$ determines that contributed by the component B's j^{th} mode. From (25), we note that the sum of these relative contribution matrices is an identity matrix.

Given a system damping matrix C_{dd} with certain desirable properties, optimal values of c^{A_i} and c^{B_j}, $(i,j) = 1,...,m$, which best satisfy equation (26) are determined via solving a minimization problem with the following cost functional J:

$$\min_{c^{A_i}, c^{B_j}} J = \frac{1}{2} \| C_{dd} - \sum_{i=1}^{i=m} c^{A_i} R_{dd}^{A_i} - \sum_{j=1}^{j=m} c^{B_j} R_{dd}^{B_j} \|_F^2 , \qquad (28)$$

where $\| \bullet \|_F^2$ is the squared Frobenius norm of the matrix concerned. The cost functional J could also be rewritten as:

$$J = \frac{1}{2} \sum_{r=1}^{r=d} \sum_{s=1}^{s=d} [E_{dd}]_{rs}^2 , \qquad (29)$$

where the (r, s) element of the error residual matrix E_{dd} is given by:

$$[E_{dd}]_{rs} = [C_{dd}]_{rs} - \sum_{i=1}^{i=m} c^{A_i} [R_{dd}^{A_i}]_{rs} - \sum_{j=1}^{j=m} c^{B_j} [R_{dd}^{B_j}]_{rs} , \qquad (30)$$

where $[\bullet]_{rs}$, $(r,s) = 1,...,d$, denotes the (r, s) element of the $d \times d$ matrix ($\bullet$). The unknown variables c^{A_i} and c^{B_j} are determined using the following optimality conditions:

$$\frac{\partial J}{\partial c^{A_i}} = \frac{\partial J}{\partial c^{B_j}} = O_{1m} , \qquad (31)$$

where $(i,j) = 1,...,m$, and O_{1m} is a $1 \times m$ null matrix. Accordingly, the optimal values of c^{A_i} and c^{B_j} are given by:

$$[c^{A_1}, \quad \cdots, \quad c^{A_m}, \quad c^{B_1}, \quad \cdots, \quad c^{B_m}]^T = Q^{-1} \times f , \qquad (32)$$

where the $2m \times 2m$ Q matrix and the $2m \times 1$ f matrix are defined as:

$$Q \triangleq \begin{bmatrix} \| R_{dd}^{A_1} R_{dd}^{A_1} \| & \cdot\cdot & \| R_{dd}^{A_1} R_{dd}^{A_m} \| & \cdot\cdot & \| R_{dd}^{A_1} R_{dd}^{B_m} \| \\ \| R_{dd}^{A_2} R_{dd}^{A_1} \| & \cdot\cdot & \| R_{dd}^{A_2} R_{dd}^{A_m} \| & \cdot\cdot & \| R_{dd}^{A_2} R_{dd}^{B_m} \| \\ \cdot\cdot & & \cdot\cdot & & \cdot\cdot \\ \| R_{dd}^{B_m} R_{dd}^{A_1} \| & \cdot\cdot & \| R_{dd}^{B_m} R_{dd}^{A_m} \| & \cdot\cdot & \| R_{dd}^{B_m} R_{dd}^{B_m} \| \end{bmatrix} , \qquad (33)$$

$$f \triangleq \begin{bmatrix} \| R_{dd}^{A_1} C_{dd} \| \\ \cdots \\ \| R_{dd}^{A_m} C_{dd} \| \\ \| R_{dd}^{B_1} C_{dd} \| \\ \cdots \\ \| R_{dd}^{B_m} C_{dd} \| \end{bmatrix} . \qquad (34)$$

In (33-34), $\| X Y \| \triangleq \sum_{i-1}^{d} \sum_{j=1}^{d} X_{ij} \times Y_{ij}$. Using a generalized form of Cauchy's inequality theorem,[11] it can be shown that the determinant of Q is always greater than zero unless the matrices $R_{dd}^{A_1},..., R_{dd}^{A_m}, R_{dd}^{B_1},...,R_{dd}^{B_m}$ are linearly dependent. In that case, a SVD procedure could be used to select a reduced set of independent modes from the original mode set. On the other hand, if the matrices $R_{dd}^{A_1},...,R_{dd}^{B_m}$ are linearly independent, the matrix Q is non-singular, and the unknowns c^{A_i} and c^{B_j}, $(i, j) = 1,...,m$ are determined using (32).

The cost functional of the above formulated optimization problem could be modified using a $d \times d$ symmetric weighting matrix W_{dd}:

$$J_W = \frac{1}{2} \sum_{r=1}^{r=d} \sum_{s=1}^{s=d} [W_{dd}]_{rs}^2 [E_{dd}]_{rs}^2 . \qquad (35)$$

If the diagonal elements of the weighting matrix W_{dd} have values that are relatively larger than the off-diagonal terms, the values of c^{A_i} and c^{B_j} computed with the modified cost functional J_W will lead to an error residual matrix E_{dd} with small diagonal terms. That is, the desired damping factors of the system modes are closely captured. But this is at the expense of having "large" off-diagonal terms in the resultant system damping matrix. The reverse is true if the diagonal elements of W_{dd} have relatively small values. One way to select a "compromised" weighting matrix will be illustrated in the sequel.

Results obtained using the above formulated unconstrained optimization problem might or might not be "physically meaningful." Situations arise in which, for a given C_{dd}, one or more computed damping terms c^{A_i} and c^{B_j}, $(i,j) = 1,...,m$, is negative. While positive values for c^{A_i} and c^{B_j} imply energy dissipation, a negative c^{A_i} (or c^{B_j}) implies energy addition, which is physically impossible. To overcome this difficulty, the following inequality constraints are added to the above formulated optimization problem:

$$1 \geq c^{A_i}/[2\omega^{A_i}] > 0, \qquad i = 1, ..., m,$$
$$1 \geq c^{B_j}/[2\omega^{B_j}] > 0, \qquad j = 1, ..., m. \tag{36}$$

The additions of these inequality constraints make it impossible to solve the optimization problem analytically. However, we can still solve it using, for example, the NPSOL package.[12] This is an optimization package that is designed to minimize a smooth function subjected to constraints (including simple bounds on the design variables and other linear/nonlinear constraints). The effectiveness of the proposed component modes damping assignment method, called ModeDamp, is established using the following example.

6.4. APPLYING MODEDAMP ON A GALILEO MODEL

The effectiveness of ModeDamp is established by applying it on a high-order finite-element model of the Galileo spacecraft. The three-body topology of the dual-spin Galileo spacecraft is illustrated in Figure 2.[2] The rotor is the largest and most flexible component represented with 243 dof's. The smaller and more rigid stator is represented with 57 dof's. Lastly, the scan platform is the smallest body idealized as rigid with 6 dof's.

---

‡If $C_{dd} = \beta_1 I_{dd} + \beta_2 \Lambda_{dd}$, where both β_1 and β_2 are positive constants, then the computed values of c^{A_i} and c^{B_j} are guaranteed to be positive.

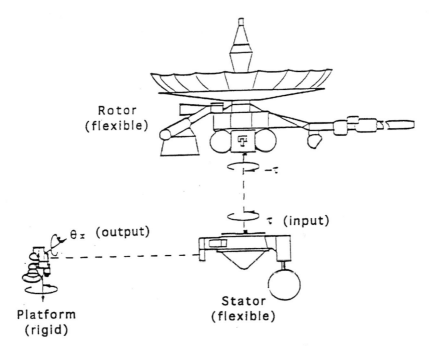

Figure 2. Galileo spacecraft model.

For the purpose of controller design, low-order system models, at all configurations of interest, and over a frequency range of 0-10 Hz are needed. To this end, we apply the MacNeal-Rubin version of the COMPARE methodology on the Galileo model. The first stage of COMPARE requires the generations of MacNeal-Rubin mode sets for all the flexible components. Following standard procedures, free interface normal modes of both the rotor and stator are first determined, and then truncated at twice the frequency of interest (20 Hz). Next, these truncated normal mode sets are each augmented with residual modes. The resultant mode sets are then used to generate reduced-order models for both the rotor and stator.

Next, the CMS-generated component models are assembled using the interface displacement compatibility relations to construct system models at several system configurations. Using the modal influence coefficient[2] as a selection criterion, important system modes are determined at system configurations with clock angles of 0, 60, 120, 180, 240, and 300 degrees. A composite mode set that encompasses important modes at all configurations is then determined. This set, with 8 rigid-body and 21 flexible modes, is indicated in Table 1.

Table 1. Frequencies of full-order system flexible modes (clock angle = 300 degrees)

Mode	Ω_{system} (Hz)	Mode	Ω_{system} (Hz)	Mode	Ω_{system} (Hz)
1	0.127	31	5.634	61	9.606[†]
2	0.864[†]	32	5.652	62	10.179
3	1.227	33	5.652	63	10.306
4	1.236[†]	34	5.660	64	10.420[†]
5	1.238	35	5.660	65	10.555[†]
6	1.479[†]	36	5.662	66	13.534[†]
7	1.522	37	5.783	67	13.990
8	1.546	38	5.785	68	16.800[†]
9	1.602	39	5.809	69	16.996
10	1.707[†]	40	5.814	70	17.122
11	1.734[†]	41	5.824	71	17.977
12	2.072[†]	42	5.824	72	18.063
13	2.341	43	5.830	73	18.083
14	2.351[†]	44	5.830	74	18.186
15	2.815[†]	45	5.833	75	18.527
16	3.073	46	5.833	76	18.688
17	3.232	47	5.834	77	19.591
18	3.233	48	5.834	78	19.593
19	3.707[†]	49	5.835	79	21.280[†]
20	4.172	50	5.835	80	27.537
21	4.977	51	5.835	81	29.919
22	5.022	52	5.839	82	31.083
23	5.231[†]	53	5.901	83	55.686
24	5.247	54	6.034	84	67.384
25	5.455	55	6.161	85	71.770[†]
26	5.455	56	6.221	86	82.129[†]
27	5.467[†]	57	6.869	87	87.522
28	5.588	58	7.056[†]	88	96.235
29	5.588	59	7.190	89	100.95
30	5.633	60	8.153[†]	90	274.45

[†] flexible system modes in the composite mode set.

The next step is to augment the composite mode set with static correction modes. For the Galileo example, two residual modes, one for an input torque about the Z-axis on the rotor side of the rotor/stator interface, and a second for an equal and opposite torque on the stator side of the interface were generated to augment the composite mode set. The enlarged mode set is then projected onto the flexible component models, and a SVD was used to remove linearly dependent modes in the projected stator mode set. The resultant reduced-order models of the rotor and stator have 29 (with 6 rigid-body) and 21 (with 8 rigid-body) modes, respectively. The assembled reduced-order system model has 44 (with 8 rigid-body) modes (see also Table 2). Note that all modes retained in the composite mode set have been captured exactly

Table 2. Reduced-order Component and System Models' Modes

Flexible Mode	ω_{stator} (Hz)	ω_{rotor} (Hz)	ω_{system} (Hz)
1	7.105	0.143	0.127
2	9.129	0.866	0.864[†]
3	10.561	1.237	1.236[†]
4	14.560	1.483	1.479[†]
5	43.172	1.728	1.707[†]
6	50.070	2.286	1.734[†]
7	63.530	2.809	2.072[†]
8	80.867	3.647	2.351[†]
9	86.989	3.996	2.815[†]
10	96.605	5.207	3.707[†]
11	166.34	5.337	4.167
12	240.18	5.994	5.231[†]
13	254.52	6.410	5.433
14		9.503	5.467[†]
15		10.291	6.150
16		10.553	6.436
17		13.536	7.056[†]
18		15.613	8.153[†]
19		29.188	9.606[†]
20		41.181	9.613
21		58.524	10.294
22		69.003	10.420[†]
23		77.722	10.555[†]
24			13.534[†]
25			13.997
26			15.860
27			16.800[†]
28			20.591
29			21.280[†]
30			28.462
31			34.316
32			41.214
33			48.011
34			58.318
35			71.770[†]
36			82.129[†]

[†] exactly captured system flexible modes (see Ω_{system} in Table 1).

in the reduced-order system model. A Bode plot comparison of the full and reduced-order models, at a clock angle of 300 degrees, shows excellent result (see Figure 3). Comparisons made at all other clock angles are equally impressive.[3,10]

Given a system damping matrix C_{dd} with certain "desirable" properties, the ModeDamp methodology is next used to determine the required component damping factors. But what are these "desirable" damping properties?

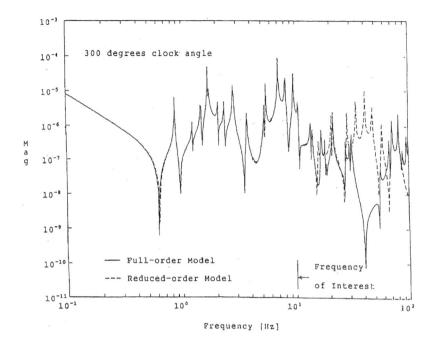

Figure 3. Bode plot comparison of the full and reduced-order models.

We might want to have the modal damping factors of the system damping matrix closely match those found experimentally. To this end, the measured damping factors of the TOPEX/POSEIDON satellite could be used as a guide:[13]

$$1.0\% \quad \text{for} \quad 5 \leq \omega \leq 8.5 \quad \text{Hz},$$
$$0.5\% \quad \text{for} \quad 8.5 < \omega \leq 11 \quad \text{Hz},$$
$$\text{and} \quad 1.0\% \quad \text{for} \quad 11 < \omega \leq 35 \quad \text{Hz}.$$

In our study, the system damping matrix C_{dd} is chosen to be diagonal with the following properties (see also Table 4):

(1) A uniform damping factor of 0.25% for all the recaptured dominant system modes that fall within the frequency range of interest (0–10 Hz). That is, $\zeta_i = 0.25\%$, for $i = 2$–10, 12, 14, and 17–19.

(2) The damping factors of all the recaptured system modes that fall outside the frequency range of interest must be at least 0.25%. That is, $\zeta_j \geq 0.25\%$ for $j = 22$–24, 27, 29, 35, and 36; and

(3) The damping factors of the remaining system modes must be larger than zero. That is, $\zeta_k > 0$ for $k = 1, 11, 13, 15-16, 20-21, 25-26, 28,$ and $30-34$.

These requirements on the system damping factors are explained as follow. The recaptured system modes are modes that make significant contribution to the system input-to-output mapping. For recaptured modes that fall within the frequency range of interest, damping factors of 0.25% are consistent with experimental data found with large space structures.[6-7] Recaptured system modes that fall outside the frequency range of interest make little contribution to the system input-to-output mapping. But damping factors larger than 0.25% can help to "damp out" transient dynamics associated with these modes. All remaining system modes can assume any damping factors that are larger than zero. The ModeDamp methodology could also be used with other appropriately selected system damping matrix.

The 36×36 weighting matrix W_{dd} in equation (35) must be carefully constructed to realize requirement [1]. First, the diagonal elements of W_{dd} are selected as follow: $W_{dd}(i, i) = \zeta_2 \omega_2 / \zeta_i \omega_i$ $(i = 2\text{-}10, 12, 14,$ and $17\text{-}19)$ and $W_{dd}(i, i) = 0$ for the remaining modes. These selections give equal weightings to all the recaptured system modes that lie within the frequency range of interest while paying no attention to those that lie outside the frequency range. All the off-diagonal elements of the ith row of W_{dd} are given the weighting $p\,W_{dd}(i, i)$. If "p" is small, our attention will be on the diagonal elements, and the desired 0.25% damping factor will be closely captured. However, the magnitudes of some off-diagonal elements in C_{dd} might be larger than the corresponding diagonal elements. The matrix C_{dd} is no longer diagonally dominant, which is not what we want. Conversely, with a large "p", we can produce a truly diagonally dominant C_{dd}, but the desired 0.25% damping factor for $i = 2\text{-}10, 12, 14,$ and $17\text{-}19$ are only captured approximately. Via a trial-and-error process, p was selected to be 0.11.

Several inequality constraints on the component damping factors must also be enforced:

$$1 \geq c_i^r / [2\omega_i^r] > 0, \qquad i = 1, ..., 23,$$
$$1 \geq c_j^s / [2\omega_j^s] > 0, \qquad j = 1, ..., 13,$$

where the superscripts "r" and "s" denote damping factors associated with the rotor and stator, respectively. Additionally, we must also enforce system-level inequality constraints in accordance with requirements [2] and [3]. These constraints are:

$$\text{diag}(U) \geq \text{diag}(\sum_{i=1}^{i=23} c_i^r R_i^r + \sum_{j=1}^{j=13} c_j^s R_j^s) > \text{diag}(L),$$

Table 3. Damping Factors of the Rotor and Stator Flexible Modes

Mode	Rotor Freq (Hz)	Rotor ζ (%)	Stator Freq (Hz)	Stator ζ (%)
1	0.143	0.4799	7.105	0.3791
2	0.866	0.2505	9.129	0.1638
3	1.237	0.2508	10.561	0.2474
4	1.483	0.2502	14.560	0.6307
5	1.728	0.4116	43.172	1.0851
6	2.286	0.3238	50.070	0.9665
7	2.809	0.2514	63.530	2.1919
8	3.647	0.2214	80.867	3.1790
9	3.996	0.7162	86.989	0.4047
10	5.207	0.2435	96.605	2.1883
11	5.337	0.3111	166.34	10.011
12	5.994	0.4924	240.18	6.8794
13	6.410	0.7881	254.52	5.4772
14	9.503	0.6807		
15	10.291	0.0070		
16	10.553	0.2503		
17	13.536	0.3060		
18	15.613	0.9075		
19	29.188	0.9720		
20	41.181	1.2991		
21	58.524	1.3902		
22	69.003	2.0869		
23	77.722	2.8904		

where U and L denote an upper and a lower bound vectors, respectively. With $U = 2 \times 1 \times [\omega_1^{sys}, ..., \omega_{36}^{sys}]^T$, the damping factors of all the system modes will be less than unity. Similarly, with $L = O_{36 \times 1}$, all system dampings will be "dissipative." To enforce requirement [2], the "zero" entries in the jth elements of L (j = 22-24, 27, 29, and 35-36) are replaced by $2 \times 0.0025 \times \omega_j^{sys}$.

The damping factors of the reduced-order rotor and stator models determined using ModeDamp are tabulated in Table 3. The resultant damping factors of the reduced-order system modes are given in Table 4. In that table, damping factors of all recaptured system modes that lie within the frequency range of interest are indeed close to 0.25%. The damping factors of modes 5 and 7, at 0.1850% and 0.1607%, respectively, are lower than 0.25%. The situation could be improved by increasing the magnitudes of the (5,5) and (7,7) elements in the weighting matrix W_{dd}. However, this will lead to a system damping matrix with large off-diagonal terms for modes 5 and 7, which is undesirable. The damping factors of all the recaptured system modes that lie outside the frequency range of interest are larger than 0.25%. This satisfy requirement [2]. The resultant C_{dd} matrix is also diagonally dominant.

Table 4. Damping Factors of the Reassembled System's Flexible Modes

Mode	ω_{sys} (Hz)	$\zeta_{sys}(\%)$ (= 0.25%)	$\zeta_{sys}(\%)$ (≥ 0.25%)	$\zeta_{sys}(\%)$ (> 0%)
1	0.127			0.4261
2	0.864†	0.2500		
3	1.236†	0.2500		
4	1.479†	0.2487		
5	1.707†	0.1850		
6	1.734†	0.2795		
7	2.072†	0.1607		
8	2.351†	0.2748		
9	2.815†	0.2503		
10	3.707†	0.2520		
11	4.167			0.6438
12	5.231†	0.2503		
13	5.433			0.3103
14	5.467†	0.2401		
15	6.150			0.1784
16	6.436			0.7833
17	7.056†	0.2332		
18	8.153†	0.2577		
19	9.606†	0.2602		
20	9.613			0.6782
21	10.294			0.0090
22	10.420†		0.2500	
23	10.555†		0.2505	
24	13.534†		0.3070	
25	13.997			0.5977
26	15.860			0.8979
27	16.800†		0.6157	
28	20.591			0.3918
29	21.280†		0.8268	
30	28.462			0.7813
31	34.316			1.0344
32	41.214			1.2990
33	48.011			1.2128
34	58.318			1.6181
35	71.770†		2.2640	
36	82.129†		2.1006	

† Exactly captured system modes.

In fact, if $r_i = \max_{j=1,\ldots,36, j \neq i} | c_{ij}/c_{ii} |$ represents the maxima of all ratios of the off-diagonal terms to the diagonal term, for mode i, then, $r_i = 0.49$, 0.17, 0.07, 0.12, 0.47, 0.07, 0.04, 0.36, 0.40, 0.02, 0.27, 0.11, 0.06, and 0.35, for i = 2-10, 12, 14, 17-19 (dominant system modes), respectively.

6.5. CONCLUDING REMARKS

To simulate the dynamic motion of articulated multi-flexible body structures, one can use multibody simulation packages such as DISCOS. To this end, one must provide appropriate reduced-order models for all the flexible components involved. The COMPARE methodology is an effective way to generate these reduced-order component models. In conjunction, we must also provide component damping matrices which when reassembled generate a system damping matrix which has certain desirable damping properties.

To determined the needed component damping factors, we establish from first principles a matrix-algebraic relation between the system's modal damping matrix and the components' modal damping matrices. A constrained optimization problem is then formulated to determine the component damping factors that best satisfy that matrix-algebraic relation. In this study, the proposed technique was "tailored" to the case when the component models were obtained by the COMPARE methodology. However, it could also be used on component models that are determined using an alternative model reduction methodology. The effectiveness of the methodology had been successfully validated using a high-order finite-element model of the Galileo spacecraft.

Acknowledgments

The research described here was conducted at Jet Propulsion Laboratory, California Institute of Technology under a contract with the National Aeronautics and Space Administration. The author wishes to thank his colleagues D. Bernard, F. Hadaegh, R. Laskin, G. Macala, S. Sirlin, J. Spanos, W. Tsuha, and M. Wette; J. Storch at the C.S. Draper Laboratory, and A. von Flotow, formerly with the Massachusetts Institute of Technology for many helpful discussions and valuable suggestions. All errors are my responsibility.

References

1. Bodley, C.S., Devers, A.D., Park, A.C., and Frisch, H.P., 1978, "A Digital Computer Program for the Dynamic Interaction Simulation of Controls and Structure (DISCOS)," *NASA Technical Paper 1219*, Vols. I and II.
2. Lee, A.Y. and Tsuha, W.S., 1994, "A Model Reduction Methodology for Articulated Multi-flexible Body Structures," *Journal of Guidance, Control, and Dynamics*, **17**(1), 69–75.
3. Lee, A.Y. and Tsuha, W.S., 1994, "A Component Model Reduction Methodology for Articulated Multi-flexible Body Structures," *Journal of Guidance, Control, and Dynamics*, **17**(4), 864–868.

4. Hasselman, T.K., 1976, "Damping Synthesis from Substructure Tests," *AIAA Journal*, **14**, 1409–1418.
5. Béliveau, J.G. and Soucy, Y., 1985, "Damping Synthesis Using Complex Substructure Modes and A Hermitian System Representation," *AIAA Journal*, **23**(12), 1952–1956.
6. Lim, K.B., Maghami, P.G., and Joshi, S.M., 1992, "Comparison of Controller Designs for an Experimental Flexible Structure," *IEEE Control Systems*, pp. 108-118.
7. Macala, G., 1984, "A Model Reduction Method for Use With Nonlinear Simulations of Flexible Multi-body Spacecraft," AIAA Paper 84-1989, *Proceedings of the AIAA/AAS Astrodynamics Conference*, Seattle, Washington.
8. Craig, R.R., Jr., 1981, *Structural Dynamics: An Introduction to Computer Methods*, John Wiley and Sons, Inc., New York.
9. Bernard, D., 1990, "A Projection and Assembly Method for Multi-body Component Model Reduction," *Journal of Guidance, Control, and Dynamics*, **13**(5).
10. Lee, A.Y., 1993, "A Component Modes Damping Assignment Methodology for Articulated, Multi-flexible Body Structures," *Journal of Guidance, Control, and Dynamics*, **16**(6), 1101–1108.
11. Hardy, G.H., Littlewood, J.E., and Pólya, G., 1959, *Inequalities*, Cambridge University Press.
12. Gill, P.E., Murray, W., Saunders, M., and Wright, M.H., 1986, "User's Guide for NPSOL (4.0): A Fortran Package for Nonlinear Programming," TR SOL 86-2, System Optimization Laboratory, Stanford University.
13. Larkin, P.A., Francis, J.T., Robinson, F.Y., and Lou, M., 1992, "Structural Development of the TOPEX/POSEIDON Satellite," *Sound and Vibration*, pp. 14-23.